CHARLES DARWIN IN AUSTRALIA

....

F.W. Nicholas & J.M. Nicholas

Charles Darwin in Australia

F.W. NICHOLAS &
J.M. NICHOLAS

WITH ILLUSTRATIONS AND ADDITIONAL
COMMENTARY FROM OTHER MEMBERS OF THE *BEAGLE'S*
COMPANY INCLUDING CONRAD MARTENS,
AUGUSTUS EARLE, CAPTAIN FITZROY, PHILIP GIDLEY KING,
AND SYMS COVINGTON

CAMBRIDGE
UNIVERSITY PRESS

CAMBRIDGE UNIVERSITY PRESS
Cambridge, New York, Melbourne, Madrid, Cape Town, Singapore, São Paulo

Cambridge University Press
477 Williamstown Road, Port Melbourne, VIC 3207, Australia

Published in the United States of America by Cambridge University Press,
New York

www.cambridge.org
Information on this title: www.cambridge.org/9780521728676

First published 1989
Paperback edition 2002
Anniversary edition 2008

Designed and typeset by Mary Mason, Mason Design
Cartography by Tony Fankhauser
Printed in China by Printplus Limited

A catalogue record for this publication is available from the British Library

National Library of Australia Cataloguing in Publication data

Nicholas, F.W.
Charles Darwin in Australia / F.W. Nicholas, J. M. Nicholas.
Anniversary 2nd ed.
9780521728676 (hbk.)
Includes index.
Bibliography.
Darwin, Charles, 1809–1882—Diaries.
Darwin, Charles, 1809–1882—Travel—Australia.
Beagle Expedition 1831–1836.
Australia—Description and travel.
Nicholas, J.M. (Janice Mary) 1944–

919.4042

ISBN 978-0-521-72867-6 hardback

To Chris and Hannah,

to the memory of Charles and Mary,

and to the anonymous donor

who made it all possible.

. . . .

A Conrad Martens watercolour showing Sydney Cove viewed from Bunkers Hill, as it appeared at the time of Darwin's visit to Australia in 1836.

FOREWORD

....

Just before midnight, the red revolving beacon of Sydney's lighthouse hove in sight, winking above the swell, welcoming the tiny ten-gun brig HMS *Beagle* to the colony of New South Wales. As the sun rose, she slipped through The Heads into Port Jackson and steered gently to the safety of Sydney Cove. It was 12 January 1836 and high summer, forty-eight years less a week since the First Fleet reached Botany Bay. Charles Darwin stepped ashore in Australia.

Half-way round the world after four years at sea, he was 'bitterly disappointed' to find no letters. 'I feel much inclined to sit down & have a good cry', he sighed, and then no doubt he did so right there in The Rocks. He was emotional: he carried Milton's *Paradise Lost* for shore-reading and could hear Handel's 'Messiah' on mountain tops as he peered down on the creation. And God was still in Darwin's antipodean heaven. For all the enigmas posed by this oddly inverted world, the ex-ordinand confessed, 'One hand has surely worked throughout the universe'.

He stayed for two months – six days sampling Sydney, eleven trekking into the Blue Mountains, a dozen in Van Diemen's Land (today's Tasmania, where he celebrated his twenty-seventh birthday), eight at King George Sound and the rest in the *Beagle* bobbing from port to port. Distracted he might be, pining for home and, frankly, a wife. 'Two years from England | pretty daughter', he jotted in a pocket pad after a family luncheon in the country. He looked forward, too, with 'a comical mixture of dread & satisfaction' to the voyage's scientific aftermath. But no thoughts of home, or seeing 'Convicts driving in their carriages & revelling in Wealth', diminished Darwin's wonder. The fauna here were beguiling, the Aborigines most 'curious' and the geology (he jotted in his pad) 'magnificent', 'astounding & unique'. To this day, Darwin's diary, notebooks, letters and publications radiate enthusiasm for Australia.

After studying these documents for decades and writing Darwin's biography with Adrian Desmond, I went to see for myself. Not even Darwin's words had prepared me for the spectacle of Terra Australis, its scents and sounds, the patterns of light and shade, the openness of spaces and of the people. One fine morning in spring 1996, the Nicholases took me into the Blue Mountains on the route of Darwin's January 1836 expedition. I had revelled in their *Charles Darwin in Australia*, poring over the old photos and the paintings by the *Beagle* artists Augustus Earle and Conrad Martens. Strangely I now found these images *did* prepare me for the country. Carried beyond Darwin's words, I knew what to look for, what to *see*, even before my hosts' running commentary brought each place to life.

Driving up through Katoomba, we drank in Darwin's bar at the Gardner's Inn, Blackheath, and walked to the site of the Weatherboard Inn where he rested. A short hike down Jamison Creek and I was stunned as the earth fell away into the vast gulf of Wentworth Falls, a 'grand Amphitheatre' to Darwin; yet I too found Govett's Leap even 'more stupendous'. Most memorable was our crossing the Nepean at Emu Ferry to reach the site of Dunheved farm, where the *Beagle*'s retired captain Philip Parker King hosted Darwin for a day and a night. The old homestead was long-gone and the land abandoned. Ignoring the signs, Frank and I slithered under the mesh fence like schoolboys and made a bee-line to the foundations where, digging with our fingers in an overgrown garden, we unearthed shards of English china. For a moment, Darwin must have felt at home here.

In England before the voyage, Darwin had longed to visit the tropics, and the scenic word-pictures in Alexander von Humboldt's *Personal Narrative* fired his imagination. But gazing on prints by Rugendas and Clarac showed him so much more, 'the infinite numbers of lianas & parasitical plants & the contrast of the flourishing trees with the dead & rotten trunks'. These subtle and intricate images prepared Darwin to *see* the New World, however much the Brazilian forests finally transcended his expectations.

This new edition of *Charles Darwin in Australia* is Humboldt, Rugendas and Clarac rolled into one for the twenty-first century Darwinian traveller – and more. Every documentary resource, every manuscript and printed text is laid out in order, with maps, and knitted together by lively commentary. Photos give a sense of then-and-now that would have astonished the young Darwin, but his 'cameramen in Australia', as the authors call them, the artists Earle, Martens and John Gould, offer the most vivid revelations. In this edition as never before, voyagers on the southern continent can see what Darwin saw, much as he saw it. All original paintings and engravings are reproduced in glorious colour.

Charles Darwin in Australia, new and improved, is not only a labour of love worthy of the Darwin bicentenary. It is also a richly textured study of historic science in-the-making. No finer, focussed account exists of the spaces, the places and the people who made up Darwin's world, and whose world he would remake.

JAMES MOORE
CAMBRIDGE, ENGLAND

CONTENTS

....

'one stands on the brink of a vast precipice. Below is the grand
bay or gulf . . . the line of cliff diverges away on each side, showing
headland, behind headland, as on a bold Sea coast.'

(FROM DARWIN'S DIARY, 17 JANUARY 1836)

PREFACE

....

2009 marks the 200th anniversary of the birth of Charles Darwin and the 150th anniversary of the publication of *On the Origin of Species by Means of Natural Selection*. This Anniversary edition of the account of Darwin's visit to Australia has been published to coincide with these commemorations.

Charles Darwin's account of his five-year voyage around the world in HMS *Beagle* proved to be very popular with the general public when it was first published in 1839. A second edition, published in 1845, was equally successful.

Included in this account is a section on Australia, which Darwin visited from January to March 1836. His descriptions of the towns of Sydney, Hobart and Albany and of his trip to Bathurst are still very readable, and provide informative glimpses of early colonial life.

By a fortunate coincidence, Darwin's shipmates on the *Beagle* during an earlier stage of the voyage included the artists Augustus Earle and Conrad Martens. Both artists spent time in Australia, and both have left behind paintings and drawings depicting what they saw. In many cases, these provide exactly contemporary illustrations of scenes described by Darwin. Earle stayed in Australia for only a few years, but is now regarded as an important colonial artist. Martens lived in Sydney from 1835 until his death in 1878, and became the most prolific and probably the best known of all colonial artists.

By the time Darwin arrived in Sydney in January 1836, Martens had already made many sketches and some paintings of Sydney, Parramatta, Penrith and the Blue Mountains, all of which were visited and described by Darwin. In fact, Darwin mentioned some of the Martens paintings in the diary that he kept during his visit. Darwin also described a panorama of Hobart that was based on paintings by Earle.

Darwin visited Martens in his Sydney studio in January 1836, and purchased two paintings. Years later, Martens wrote to Darwin and sent another painting.

As broad as their interests were, neither Martens nor Earle was particularly active in illustrating Australia's animals, and so we had to search elsewhere for contemporary illustrations of the native animals seen by Darwin. A small number were illustrated in *The Zoology of the Voyage of H.M.S. Beagle*, and all of these plates have been reproduced in the present book. But many of the Australian animals seen by Darwin were not included in the *Zoology*.

Fortunately, in our search for suitable illustrations, we have had to look no further than those of John Gould, whom Darwin knew well. Having already established his reputation as an illustrator and cataloguer of birds, and as a talented naturalist, Gould was an obvious choice when Darwin was searching for someone to prepare the bird volume of *The Zoology of the Voyage of H.M.S. Beagle*. As soon as he had completed this job for Darwin, Gould went to Australia in May 1838 to collect specimens for his next work, *The Birds of Australia*, which eventually ran to seven volumes published between 1840 and 1848, plus an 1869 supplement, containing in total 683 beautiful coloured plates. Seeing no reason to stop at birds, Gould also collected information on other Australian animals, which later (1845 to 1863) gave rise to *The Mammals of Australia*, containing 182 equally impressive coloured plates. Since Gould's visit to Australia followed so closely behind Darwin's, and in view of their collaboration in the *Beagle* bird volume, it seemed fitting for some of Gould's plates to be reproduced in the present book, as illustrations of the animals seen by Darwin during his Australian visit.

In effect, then, the artists Martens, Earle and Gould have been used in this book as Darwin's cameramen in Australia, providing a pleasing, if at times romantic, substitute for the photographs that Darwin would undoubtedly have taken, had a camera been suitably portable at that time.

In retelling the story of Darwin's visit to Australia, we have attempted to gather together all relevant material. Our aim has been to produce a book based primarily on Darwin's own account, illustrated wherever possible by Martens, Earle, and Gould. In the case of Martens, this has provided an opportunity to publish some of his early sketches for the first time.

In collating Darwin's account of his visit, six major sources were available to us.

The first is the rough notes that he made in the field. These were written in a pocket notebook (known variously as notebook 1.3, notebook 14, or the Sydney–Mauritius notebook) that he carried with him during his trip to Bathurst; and on assorted sheets of paper that he used during his visits to Hobart and King George Sound. The notebook is now held in Down House[1], while the other field notes are held in the Darwin Archive at the Cambridge University Library. In collaboration with Dr Gordon Chancellor, of the City Museum and Art Gallery, Peterborough, we have prepared a complete transcript of the Australian section of the notebook. In addition, Darwin's Hobart field notes have been transcribed by Dr Chancellor;

and the notes made at King George Sound have been transcribed by us and by Dr Patrick Armstrong of the School of Earth and Geographical Sciences, University of Western Australia.

The second source is the so-called geological and zoological diaries, which contain specialist material not included in the field notes. Like the field notes, these diaries are held in the Darwin Archive at the Cambridge University Library. The Australian sections of these diaries have been transcribed in part by other workers, and in whole by us.

In the present book, extensive use has been made of the information contained in the field notes and the geological and zoological diaries, and selected entries have been incorporated in the text.[2]

The third source is the personal diary kept by Darwin during the voyage; we shall call it the Diary.[3] It was based on the entries made in the Sydney/Mauritius notebook and the other field notes. The Diary was used by Darwin as the basis for his contribution to the official record of the first and second *Beagle* voyages, which was published in May 1839[4] in three volumes, under a rather long title that we shall abbreviate to *Narrative* (see list below for the full title). Darwin's contribution was the third volume, which was given the simple title of *Journal and Remarks*. This volume proved to be so popular that it was republished three months later[5] as a separate book entitled *Journal of Researches into the Geology and Natural History of the various countries visited by H.M.S. Beagle, under the command of Captain FitzRoy, R.N., from 1832 to 1836*. Parts of the Diary were omitted from the 1839 edition, and some information not in the Diary was included. Thus the 1839 *Journal* becomes a fourth source of information on Darwin's visit to Australia.

In 1845 a second edition of Darwin's volume appeared, with 'Natural History' and 'Geology' having changed places in the title. Since it omitted some sections of the 1839 *Journal* and also introduced new material, the 1845 *Journal* is a fifth source for the present book.

Darwin's ideas changed considerably between 1836 and 1845, and it is interesting to examine these changes in the Australian section of the Diary, the 1839 *Journal*, and the 1845 *Journal*. With this in mind, the basis for Darwin's account in the present book is the Australian section of the Diary, which is reproduced here in full. Where necessary, the Diary is supplemented by a presentation and discussion of important alterations made in 1839 and 1845.

The Diary entries are from a new transcription made by the authors from the facsimile edition published by Genesis Publications (see list below). Since the Diary is in essence a rough first draft of the 1839 *Journal*, from time to time it contains incomplete sentences and misspellings. It is, however, a remarkably elegant first draft, and it captures the immediacy of Darwin's reactions to Australia in a way that the published versions of 1839 and 1845 cannot hope to emulate.

The final major source of Darwin material used in this book is the letters that he wrote during his visit to Australia, published in the first volume of *Correspondence* (see list below).

In addition to Darwin, both Captain FitzRoy and Syms Covington (Darwin's servant, who later settled in Australia) kept diaries during the voyage. Unfortunately, all but a small section of FitzRoy's diary is lost. However, FitzRoy's official account of the *Beagle*'s visit to Australia, which was presumably derived from his diary, was published in volume II of the *Narrative*. Covington's diary has survived and is available online.[6] Both FitzRoy's journal and Covington's diary have been used as additional sources for the present book. In addition, we have reproduced relevant parts of the letters written by Darwin to Covington over a period of many years after both men had settled with their families; Darwin at Down House in England, and Covington in Australia. These are currently being published in subsequent volumes of *Correspondence*.

As explained in the first chapter, other persons enter the story at various stages, including members of the King and Macarthur families, and other influential figures in the colony at that time. Material relating to these individuals, some of it in the form of unpublished manuscripts or letters to and from Darwin, has been included where relevant.

The nineteen years that have intervened since the preparation of the first edition have seen the publication of several works that are relevant – directly or indirectly – to the subject of this book. These include Richard Keynes' *Charles Darwin's Beagle Diary* (Cambridge University Press, Cambridge, 1988), Adrian Desmond's and James Moore's *Darwin* (Michael Joseph, London, 1991), Janet Browne's two-volume biography *Charles Darwin: Voyaging* and *Charles Darwin: The Power of Place* (Jonathan Cape, London, 1995, 2002), Roger McDonald's *Mr Darwin's Shooter* (Knopf, Sydney, 1998) [a novel about Syms Covington], Ann Moyal's *Platypus* (Allen and Unwin, Sydney, 2001) and *The Web of Science* (Australian Scholarly Publishing, Melbourne, 2003), Dava Sobel's *Longitude* (Walker, 1995), Susanna de Vries-Evans' *Conrad Martens on the Beagle and in Australia* (Pandanus Press, Brisbane, 1993), Elizabeth Ellis' *Conrad Martens: Life and Art* (State Library of NSW Press, Sydney, 1994), Marsden Horden's *Mariners are Warned!: John Lort Stokes and H.M.S. Beagle in Australia 1837–1843* (Miegunyah Press, Melbourne, 1989) and *King of the Australian Coast: the Work of Phillip Parker King in the Mermaid and Bathurst 1817–1822* (Miegunyah Press, Melbourne, 1997), John Pickett's and Dave Alder's *Layers of Time: the Blue Mountains and their Geology*

(Geological Survey of New South Wales, Sydney, 1997), and David Branagan's and Gordon Packham's *Field Geology of New South Wales, 3rd edn* (NSW Department of Mineral Resources, Sydney, 2000). In addition, Max Banks and David Leaman have published a detailed annotation of, and commentary on, Darwin's Hobart geology field notes (*Papers and Proceedings of the Royal Society of Tasmania* 133 (1999): 29–50); and Kenneth Smith has published additional material on Darwin's insects (*Archives of Natural History* 23 (1996): 279–86). Each of these publications adds something to parts of our story, but they have necessitated only a few minor changes in the text, the most important being some alterations to Darwin's Hobart itinerary in chapter 4. In preparing this new edition, we have also taken the opportunity to update information on the animal specimens collected in Australia during Darwin's visit (following recent research into original sources), and on some of the sites visited by Darwin. We have also made brief mention of 'Darwin's tortoise'.

By far the biggest revolution to have occurred since publication of the first edition has been the creation of internet resources, in particular the Darwin Correspondence Project (http://www.darwinproject.ac.uk/) and The Complete Works of Charles Darwin Online (http://darwin-online.org.uk/). Almost all of the Darwin resources used in the preparation of this book are now freely available from one or other of these sites. Other invaluable internet resources that have become available are The Australian Faunal Directory (http://www.environment.gov.au/biodiversity/abrs/online-resources/fauna/afd/), the EPBC Act List of Threatened Fauna (http://www.environment.gov.au/cgi-bin/sprat/public/publicthreatenedlist.pl?wanted=fauna) and the Western Australian Flora (http://florabase.calm.wa.gov.au/); all of which we have used in revising the text.

In summary, as with the first edition, the aim of this book is to present the story of Charles Darwin's visit to Australia, and to show how that story provides new perspectives not only on Darwin and the colony, but also on Martens, Earle, Covington, and the substantial number of other people who, in one way or another, were involved with his visit.

SYDNEY, MARCH 2008

ACKNOWLEDGEMENTS

....

We especially acknowledge the anonymous donor, without whose help the first edition (and hence this edition) would never have been published.

We wish to reiterate our thanks to all those people and organizations acknowledged in the first edition, for enabling this book to come into being, namely the Royal College of Surgeons of England, the Royal Society, the Utah Foundation, Peter Colbourne, Professor Sir Geoffrey Slaney, Professor William Hill, Sir Walter Bodmer, Henry Bartlett, Professor John Ward, Stephen Harrison, Peter Gautrey, Philip Titheradge, Dr Gordon Chancellor, Arthur Easton, Elizabeth Ellis, Suzanne Mourot, Margaret Medcalf, David Whiteford, Stephen Howell, Robin South, members of the Merimbula-Imlay Historical Society (Mrs Bartlett, Shirley Bazley, Elva Beatty, Betty Ferguson, Olive Robertson, Mrs B. Sirl, and Lilian Weston), Dr Hal Cogger, Phil Coleman, Dr John Paxton, Professor Tony Underwood, Dr Woody Horning, Dr Peter Stanbury, Merrilee Baglin, Dr Max Banks, Dr David Branagan, Professor Tony Larkum, Dr J. R. Paxton, Professor Chris Moran, Dr Robin Pellew, Dr Robin Derricourt, Tony Bishop and Denis French.[1]

For this Anniversary edition, we are grateful for the help and advice received from Alison Ward, Mary Casey, Diana Simpkins, Ashley Hay, Bethany Wilson, Sarah Murphy, Joy Lefroy, Jennifer Broomhead, Shirley Bazley, Adrian Desmond, Jonathan Usmar, Dr Glenn Shea and Associate Professor Ian Jack. We are particularly indebted to James Moore for his inspiration and for raising our spirits during the pre-anniversary year. We are especially grateful to the staff of Cambridge University Press for their encouragement, enthusiasm and professionalism; in particular we thank Pauline de Laveaux and Jodie Howell. We would also like to thank the cartographer, Tony Fankhauser. We acknowledge the scholarship of Drs Max Banks and David Leaman and express our sincere gratitude for their annotations of, and commentary on, the Hobart geological field notes which have necessitated several corrections in chapter 4, in comparison with the first edition. We are also very grateful to Dr Banks for his invaluable feedback on a draft of chapter 4.

Finally, we thank the many institutions and individuals mentioned in the detailed list of illustrations on p. 215, for permission to reproduce their images.

....

LIST OF KEY REFERENCES AND ABBREVIATIONS USED FOR THEM

....

Correspondence: *The Correspondence of Charles Darwin*. Cambridge University Press, Cambridge. (For full details of volumes published, and of letters still to be published, see http://www.darwinproject.ac.uk/).

Diary: the manuscript diary kept by Darwin during the voyage. The original is in Down House.[1]

Narrative: FitzRoy, R. (ed.), *Narrative of the Surveying Voyages of His Majesty's Ships Adventure and Beagle between the years 1826 and 1836, describing their Examination of the Southern Shores of South America, and the Beagle's Circumnavigation of the Globe*. In 3 vols, plus an appendix to vol. II. Henry Colburn, London, 1839.[2]

Vol. I: *Proceedings of the First Expedition, 1826–1830, under the command of Captain P. Parker King, R.N., F.R.S.*

Vol. II: *Proceedings of the Second Expedition, 1831–1836, under the command of Captain Robert Fitz-Roy, R.N.*

Vol. III: *Journal and Remarks, 1832–1836, by Charles Darwin.* (See following entry concerning the 1839 *Journal*.)

1839 *Journal*: Darwin, C. R. *Journal and Remarks, 1832–1836*. Henry Colburn, London, 1839. (This book is vol. III of the *Narrative*; see above. Three months after publication of the *Narrative*, vol. III was reprinted as a separate book entitled *Journal of Researches into the Geology and Natural History of the Various Countries visited by H.M.S. Beagle, under the command of Captain FitzRoy, R.N., from 1832 to 1836* [Henry Colburn, London]) (see Preface note 5).

1845 *Journal*: Darwin, C. R. *Journal of Researches into the Natural History and Geology of the Countries visited during the Voyage of H.M.S. Beagle round the World, under the command of Capt. FitzRoy, R.N.*, 2nd edn, corrected, with additions. John Murray, London, 1845.[3]

....

Coral Reefs: Darwin, C. R. *The Structure and Distribution of Coral Reefs. Being the first part of the Geology of the Voyage of the Beagle, under the command of Capt FitzRoy, R.N., during the Years 1832 to 1836.* Smith, Elder and Co., London, 1842.[4]

Volcanic Islands: Darwin, C. R. *Geological Observations on the Volcanic Islands visited during the Voyage of H.M.S. Beagle, together with some brief Notices of the Geology of Australia and the Cape of Good Hope. Being the second part of the Geology of the Voyage of the Beagle, under the command of Capt FitzRoy, R.N., during the years 1832 to 1836.* Smith, Elder and Co., London, 1844.[5]

South America: Darwin, C. R. *Geological Observations on South America. Being the third part of the Geology of the Voyage of the Beagle, under the command of Capt FitzRoy, R.N., during the years 1832 to 1836.* Smith, Elder and Co., London, 1846.[6]

Zoology: Darwin, C. R. (ed.), *The Zoology of the Voyage of H.M.S. Beagle, under the command of Captain FitzRoy, R.N., during the Years 1832 to 1836.* Smith, Elder and Co., London, 1838-1845.[7]

Part I. *Fossil Mammalia*, by Richard Owen, in 4 numbers, 1838–1840.

Part II. *Living Mammalia*, by George R. Waterhouse, in 4 numbers, 1838–1839.

Part III. *Birds*, by John Gould (and G. R. Gray), in 5 numbers, 1838–1841.

Part IV. *Fish*, by Leonard Jenyns, in 4 numbers, 1840–1842.

Part V. *Reptiles (and Amphibia)*, by Thomas Bell, in 2 numbers, 1842–1843.

A NOTE ON THE TRANSCRIPTIONS

....

The following chapters present the *whole* of the Australian section of Darwin's 1836 Diary, *in the exact order in which it was originally written*. It is necessary to emphasise this point because with so much intervening text supplied by us, it may appear that the Diary extracts have been edited so as to fit more conveniently with the accompanying text. In fact, the intervening text has been arranged specifically so as to complement the contents of the Diary.

In order to clearly distinguish all transcriptions and other quoted material from the accompanying text, the former have been indented and set in italics. Wherever possible, the source is indicated in the accompanying text. When this would have disturbed the flow of the text, the source is indicated in round brackets at the end of the quoted passage.

In transcribing the Australian section of the Diary, as well as the various other manuscript sources presented in this book, we have followed the conventions and practices used by the editorial team assembling the *Correspondence*, as outlined in 'A Note on Editorial Policy', which can be found at the beginning of each volume of that work.

In particular, certain points that would normally be interpreted as full stops have been changed to commas or omitted when their position indicates that they could not have been intended as full stops; new paragraphs have been started in long passages of text whenever the subject matter clearly changes; misspellings have been preserved ('sic' has been used as sparingly as possible); and in the one instance where the position of a line-break is important, it is indicated by a vertical bar. The only point where our practice differs from that of the editors of the *Correspondence* concerns underlined words: we have retained the underline.

The most difficult task in transcribing the Diary was in deciding which of the many alterations and additions that appear in the manuscript were made at the time the Diary was first written, and which were made later, when the Diary manuscript was being marked up for publication as the 1839 *Journal*. On close examination, it appeared to us that most of the alterations and additions fall into the second category, in which case they have been omitted from the Diary transcription. If such changes are important, they have been specifically mentioned in the accompanying text.

All passages from letters that appear in the *Correspondence* have been copied exactly from that source, except for the occasional insertion of an additional new paragraph. We gratefully acknowledge the editors of the *Correspondence* for their permission to reprint these passages.

As in the *Correspondence*, the following abbreviations and conventions have been used when quoting manuscripts in the normal (Roman) sections of the present book:

del	deleted
illeg	illegible
ins	inserted

[some text]	'some text' is an editorial insertion
[some text/	'some text' is the conjectured reading of an ambiguous word or passage
[some text]	'some text' is a description of a word or passage that cannot be transcribed, e.g. *illeg*.
\<some text\>	'some text' is a suggested reading for a destroyed word or passage
\<some text\>	'some text' is a description of a destroyed word or passage, e.g. *signature excised*

For the quoted text that appears in italics in the present book, we have reversed the use of italics and roman in editorial markings:

del	deleted
illeg	illegible
ins	inserted

[some text]	'some text' is an editorial insertion
[some text]	'some text' is the conjectured reading of an ambiguous word or passage
[some text]	'some text' is a description of a word or passage that cannot be transcribed, e.g. illeg.
\<some text\>	'some text' is a suggested reading for a destroyed word or passage
\<some text\>	'some text' is a description of a destroyed word or passage, e.g. signature excised

INTRODUCTION

IN DECEMBER 1831, THE ROYAL NAVY'S HMS *BEAGLE* LEFT

ENGLAND ON ITS SECOND SURVEYING VOYAGE.

ON BOARD WAS A YOUNG MAN CALLED

CHARLES DARWIN. DURING THE VOYAGE, HE

TOOK EVERY OPPORTUNITY TO EXAMINE THE

GEOLOGICAL FORMATIONS AND THE MYRIAD

FORMS OF PLANTS AND ANIMALS, BOTH LIVING

AND FOSSIL, IN THE DIFFERENT PARTS OF THE

WORLD VISITED BY THE *BEAGLE*.

The many observations made by Darwin during the voyage led him to question conventional wisdom on the origin of species, and sowed the seeds for his thinking about evolution. Combining his *Beagle* observations with masses of information collected after his return to England, Darwin gradually developed an idea as to how evolution could have occurred, and in 1859 he published his revolutionary book, *On the Origin of Species by Means of Natural Selection*. By providing Darwin with the initial impetus for the development of his far-reaching ideas, the *Beagle*'s voyage has become an important event in world history.

During this voyage, the *Beagle* visited Australia, giving Darwin an opportunity to examine and explore the infant colony. Owing mainly to his own understated account of the visit in his published *Journal* (1839 and 1845), the general view has emerged that Darwin did and saw nothing of importance in Australia; that the visit was of no consequence. However, examination of all the relevant material, much of it unpublished, reveals that he was actually very active and observant during his visit, that he collected numerous specimens of animals and rocks, and that he made a number of observations that played a role in the development of his ideas on evolution.

Darwin's Australian visit is also important because it provides a view of the colony through the eyes of someone who at the time was a young and unknown naturalist, but who has since become a leading figure in the history of science.

In addition, the story of the *Beagle*'s visit to Australia provides a focus on the lives of a number of people who are now remembered for their prominent achievements in science, arts, and politics.

Among those on board the *Beagle* when it visited Australia was its commanding officer, Captain Robert FitzRoy, who later became Governor of neighbouring New Zealand, and who is now remembered as the father of weather forecasting; the person who, among other things, originated the synoptic chart and the publication of daily weather forecasts in newspapers. There was also Lieutenant John Wickham, who commanded the *Beagle* in the initial stages of its third surveying voyage, and who later settled in Australia, becoming magistrate and then Government Resident in Moreton Bay, which is now the city of Brisbane. There was Midshipman Philip Gidley King (Jnr), grandson of a former Governor of the colony, who had been born near Sydney, and who later became a leading figure in the political life of the colony, as a member of the Legislative Council, and close friend of Sir Henry Parkes. And there was Syms Covington, Darwin's servant, who emigrated to the south coast of New South Wales, and is now remembered chiefly for the letters that survive from the correspondence he maintained with Darwin for many years after emigrating.

Living in the colony at the time of the *Beagle*'s visit was Phillip Parker King,[1] the Australian-born commander of the *Beagle*'s first surveying voyage, father of the

Beagle's midshipman, and one of the Royal Navy's most capable surveyors. Another local resident was Conrad Martens, formerly a shipmate on the *Beagle*, and one of Australia's best known colonial artists. Martens had been employed on the *Beagle* following the indisposition of the *Beagle*'s original artist, Augustus Earle, who also has a place in the art history of Australia. Others involved with the *Beagle*'s visit to Australia were Hannibal Macarthur, prominent landowner, member of the Legislative Council, and chairman of the Bank of Australia; Major Thomas Mitchell, explorer and Surveyor-General of New South Wales; Alfred Stephen, Solicitor-General of Van Diemen's Land and later Chief Justice of New South Wales; George Frankland, Surveyor-General of Van Diemen's Land; and Sir Richard Spencer, Government Resident at King George Sound.

The story of the *Beagle*'s visit to Australia gives a fresh perspective on these people by presenting them in the context of their interactions with Charles Darwin.

In order to appreciate these interactions, we first need to understand a little of the background to the *Beagle*'s voyage.

In 1825, the Admiralty commissioned two vessels, the *Adventure* and the *Beagle*, to survey the southern coasts of South America. In command of the *Adventure* and of the whole expedition was Captain Phillip Parker King, whose four surveying voyages in Australian waters from 1817 to 1822 had established his reputation in the Royal Navy, and earned him a Fellowship of the Royal Society at the relatively young age of 32.[2] The South American survey voyage commenced in May 1826, and achieved notable success. However, the often harsh environment took its toll: the commander

Captain Robert FitzRoy at around 30 years of age. A sketch made by Philip Gidley King (Jnr) just prior to or during the *Beagle*'s visit to Sydney.

of the *Beagle*, Captain Pringle Stokes, committed suicide at the southern-most tip of the mainland in August 1828. The *Adventure* and *Beagle* returned to Rio de Janeiro, where Robert FitzRoy, a twenty-three year old lieutenant, was appointed as Stokes's replacement. During the final two years of the surveying voyage, FitzRoy experienced the harsh environment of Tierra del Fuego at first hand – at one stage conducting surveys in an open boat for thirty-three days in the middle of winter – but proved himself to be equal to the task.[3]

On the return journey to England, the *Beagle* had more than its usual ship's company on board: there were four natives from Tierra del Fuego as well. These Fuegians had been taken hostage by FitzRoy for the return of a boat stolen by their fellow country-men. Having failed to find the boat, but having decided that some good might come from exposing his captives to English 'civilization', FitzRoy decided to take the Fuegians home with him, and after some education, to return them to their homeland.[4]

In embarking on this plan, FitzRoy believed that another surveying voyage would soon be sent to the southern coasts of South America. But the Admiralty had no such intention, and eight months after arriving in England, FitzRoy took a year's leave of absence and engaged a private vessel at his own expense in order to return his three surviving charges (one having died) to Tierra del Fuego. Before the vessel could depart, however, an influential uncle interceded with the Admiralty, and FitzRoy was commissioned to take the *Beagle* on a second surveying voyage.[5]

Such were the somewhat unusual circumstances that led to the *Beagle*'s second voyage.

Having taken the decision, however unwillingly, the Admiralty determined to make the best of it, and the newly appointed Hydrographer to the Navy, Captain Francis Beaufort, issued FitzRoy with a long list of instructions.[6] Not only was he to complete the surveys of the southern coast of South America and of the Falkland Islands, he was also to circumnavigate the world, in order to obtain an unbroken chain of longitudes, all determined by the one set of chronometers (nautical timekeepers), at as many places as possible. Since the use of chronometers for the determination of longitude is a recurring theme in the story of the *Beagle*'s visit to Australia, it is necessary to provide a little background information on the method.[7]

The technique of determining longitude by chronometer was relatively new at that time. For centuries, sailors had been able to obtain accurate determinations of latitude simply by measuring with a sextant the height of the sun above the horizon at noon local time, i.e. when the sun was at its zenith (highest point). But the determination of longitude had remained an unsolved problem for far too long. So much so that in 1715, the English Government had offered prize money of up to £20 000 to anyone who could invent a method of determining longitude.

Since the earth revolves once every 24 hours, and since its circumference had been arbitrarily divided into 360 degrees of longitude, it followed that each degree of longitude was equivalent to one 360th of twenty-four hours, which equals four minutes. Thus if the sun reached its zenith at one location eight minutes later than at another location, then those two locations must be separated by two degrees of longitude. As soon as an arbitrary reference point of zero degrees longitude had been agreed upon (the English chose the Greenwich Observatory), the problem of determining longitude became a matter of determining the difference between time at Greenwich and the local time at the place whose longitude was to be determined.

Soon after the announcement of the reward, it became evident that the most likely way to win it would be to invent a clock (chronometer) that would keep accurate time (Greenwich time) despite being subjected to extreme changes of temperature and atmospheric conditions during long voyages in tossing ships.

After a lifetime devoted to this cause, John Harrison, a carpenter from the Lincolnshire village of Barrow, with no formal training in clock making, was grudgingly rewarded by the Government in 1773. Captain Cook was the first to take a chronometer on a long surveying voyage, using a copy by Larcum Kendall of Harrison's fourth instrument with great success during his second and third voyages. Captain Bligh used this same chronometer on the ill-fated voyage of the *Bounty*.

In 1818, the Admiralty began issuing chronometers to its ships, and by the time of the *Beagle*'s first surveying voyage, King was able to take twelve with him on the *Adventure*, and a further three on the *Beagle*.[8] The virtue in taking more than one instrument lay in the fact that not all of the instruments were equally reliable, and not all could be guaranteed to continue functioning during a long voyage.

For the *Beagle*'s second voyage, FitzRoy collected together no fewer than twenty-two chronometers.[9] As Captain Beaufort said in his instructions to FitzRoy, 'Few vessels will have ever left this country with a better set of chronometers, both public and private, than the Beagle'. With such an unprecedented battery of instruments, Beaufort hoped that FitzRoy would be able to resolve the many conflicting longitude determinations that still then existed.

Captain Francis Beaufort, who, as hydrographer to the Royal Navy, issued the instructions for the *Beagle*'s second surveying voyage.

Once the South American surveys were completed, the *Beagle*'s chronometers were to be put to further use, in obtaining longitude determinations at various islands across the Pacific Ocean, from the west coast of South America to Port Jackson on the east coast of Australia. Having arrived in Port Jackson, FitzRoy was instructed to check his chronometer measurements against those of the observatory at Parramatta,

15 miles (24 kilometres) west of Sydney, which was then regarded as 'being absolutely determined in longitude'. As Beaufort noted, if this were done, then 'all those intervening islands will become standard points to which future casual voyagers will be able to refer their discoveries or correct their chronometers'. Depending on the time of year, FitzRoy was then to check the longitude at Hobart in Van Diemen's Land, and at the settlement of King George Sound, on the south-western tip of Australia. He was then to proceed home via the Keeling Islands, Mauritius, the Cape of Good Hope, and various islands in the Atlantic Ocean, using his chronometers to check the longitude at each port of call.

FitzRoy followed these instructions very closely, and thereby obtained, as he stated in his report to the Admiralty, 'a connected chain of meridian distances around the globe, the first that has ever been completed, or even attempted, by means of chronometers alone'.[10]

The reason for the *Beagle*'s visit to Australia is now clear: the settlements of Sydney, Hobart, and King George Sound were regarded as important reference points in the global chain of longitude determinations.

There is one more aspect of the instructions issued to FitzRoy that merits some attention. After describing how the surveys should be carried out and the chain of longitudes determined, and then giving some general guidelines regarding tides, trade winds, and astronomical observations, Beaufort inserted two paragraphs describing how meteorological records should be kept.[11]

This was a subject close to Beaufort's heart. In 1806 he had developed a scale of wind force and a weather code by which meteorological observations could be recorded easily in a standard manner. But it was not until he became Hydrographer to the Admiralty in 1829 that he was in a position to have his wind scale and weather code adopted by others. In fact, the first time they were used officially was when FitzRoy commenced his log on the *Beagle* in December 1831. Beaufort could not have chosen a more suitable captain to test his wind scale and weather code: FitzRoy was already very interested in meteorology, and in later life he was instrumental in establishing the world's first regular weather forecasting service. Thanks to Beaufort's instructions, and FitzRoy's dutiful following of them, we have a complete picture of the weather during every day of the *Beagle*'s voyage, including the time spent in Australia. Beaufort's wind scale proved so successful that it was adopted internationally in 1854, and is still in use today.[12]

During his preparation for the voyage, FitzRoy was aware of the harsh conditions that would be encountered in the waters off the southern coast of South America, which had led his predecessor to suicide. With this apparently in mind, he decided that it would be very useful to have a civilian companion on board; someone

FIGURES

TO DENOTE THE FORCE OF THE WIND.

0 Calm.

1 Light Air . . . . Or just sufficient to give steerage way.

2 Light Breeze . . . . ⎫ Or that in which a man- ⎫ 1 to 2 knots.
3 Gentle Breeze . . ⎬ of-war, with all sail set, ⎬ 3 to 4 knots.
4 Moderate Breeze ⎭ and clean full, would go in smooth water from ⎭ 5 to 6 knots.

5 Fresh Breeze . . . . ⎫ ⎫ Royals, &c.
6 Strong Breeze . . . . ⎪ Or that to which a well- ⎪ Single-reefed topsails and top-gall. sails.
7 Moderate Gale . . ⎬ conditioned man-of- ⎬ Double-reefed top-sails, jib, &c.
8 Fresh Gale . . . . . . ⎪ war could just carry in chase, full and by ⎪ Treble-reefed top-sails, &c.
9 Strong Gale . . . . . . ⎭ ⎭ Close-reefed topsails and courses.

10 Whole Gale . . . .Or that with which she could scarcely bear close-reefed main-topsail and reefed fore-sail.

11 Storm . . . . .Or that which would reduce her to storm stay-sails.

12 Hurricane . . . . . .Or that which no canvass could withstand.

LETTERS

TO DENOTE THE STATE OF THE WEATHER.

b Blue Sky ; (whether clear, or hazy, atmosphere).
c Clouds ; (detached passing clouds).
d Drizzling Rain.
f Foggy——f Thick fog.
g Gloomy (dark weather).
h Hail.
l Lightning.
m Misty (hazy atmosphere).
o Overcast (or the whole sky covered with thick clouds).
p Passing (temporary showers).
q Squally.
r Rain (continued rain).
s Snow.
t Thunder.
u Ugly (threatening appearances).
v Visible (clear atmosphere).
w Wet Dew.
. Under any letter, indicates an extraordinary degree.

By the combination of these letters, all the ordinary phenomena of the weather may be expressed with facility and brevity.

Examples :—Bcm, Blue sky, with passing clouds, and a hazy atmosphere.

Gv, Gloomy dark weather, but distant objects remarkably visible.

Qpdlt, Very hard squalls, with passing showers of drizzle, and accompanied by lightning with very heavy thunder.

The first published versions of Beaufort's wind scale and weather code, as they appeared in FitzRoy's volume of the *Narrative*.

Day.	Hour.	Winds.	Force	Weather.	Sympr.	Barom.	Attd. Ther.	Temp. Air.	Temp. Water.	LOCALITY.
ABSTRACT OF METEOROLOGICAL JOURNAL.										51
JANUARY, 1836.					Inches.	Inches.	°	°	°	Lat. S. Long. W.
12	Noon.	E.	2	b c		29·99	71·5		68·5 70·5 70·5	Sydney Cove.
13	9 A.M.	S.	1	c g t l p	30·19	30·04	71·5	63		..
14	..	S.S.W.	2	c g p	30·20	30·03	72	64		..
15	..	N.	2	b c g	29·87	29·79	73·5	70		..
16	..	S.	2	b c	29·95	29·81	69	66		..
17	..	W.S.W.	4	b c	30·01	29·84	68·5	65		..
18	..	VBLE.	1	c g	30·21	30·08	71·5	69		..
19	..	N.E.	1	c g	30·25	30·12	72	69		..
20	..	..	1	b c	29·86	29·84	74	72		..
21	..	VBLE.	1	b c g	29·83	29·80	75	72·5		..
22	..	E.N.E.	2	b c	30·15	30·06	76	72		..
23	..	N.E.	5	b c g	30·37	30·22	73·5	72		..
24	..	N.W.	2	b c	29·97	29·96	74	72		..
25	..	S.E.	4	c g p d	30·40	30·16	72	63		..
26	6 A.M.	S.S.W.	1	b c	30·56			58·5		..
27	9 A.M.	N.N.W.	1	b c g	30·69	30·43	71·5	64		..
28	..	W.N.W.	1	b c g p	30·63	30·41	71·5	63		..
29	..	N.E. by E.	2	b c	30·37	30·25	72·5	67		..
30	Noon.	N.E.	4	b c	30·21	30·16	74	70	71 70·5 69·5 68	Port Jackson.
31	10 A.M.	N.	4	o m	30·24	30·18	73·5	70	68 67·5	36·32 151·17

Daily weather records collected by FitzRoy, using Beaufort's wind scale and weather code, covering the period of the *Beagle*'s stay in Sydney.

with whom he could share his meals and his worries; someone who would be outside the normal pyramid of command, at the top of which the captain by necessity had to lead a somewhat isolated existence. Since FitzRoy certainly considered himself to be a scientist (he was later to become a Fellow of the Royal Society), and since the voyage was considered to be a scientific one, it made good sense to try to find a scientist who could fill the companion's role. Moreover, since a civilian companion would have no official duties on board ship, he would be free to pursue scientific interests as the need arose, thereby maximising the chances of useful discoveries being made.[13]

In FitzRoy's own words:

> Anxious that no opportunity of collecting useful information, during the voyage, should be lost; I proposed to the Hydrographer that some well-educated and scientific person should be sought for who would willingly share such accommodations as I had to offer, in order to profit by the opportunity of visiting distant countries yet little known. Captain Beaufort approved of the suggestion, and wrote to Professor Peacock, of Cambridge, who consulted with a friend, Professor Henslow, and he named Mr. Charles Darwin, grandson of Dr. Darwin the poet, as a young man of promising ability, extremely fond of geology, and indeed all branches of natural history. In consequence an offer was made to Mr. Darwin to be my guest on board, which he accepted conditionally; permission was obtained for his embarkation, and an order given by the Admiralty that he should be borne on the ship's books for provisions. The conditions asked by Mr. Darwin were, that he should be at liberty to leave the Beagle and retire from the Expedition when he thought proper, and that he should pay a fair share of the expenses of my table.[14]

Charles Darwin was born in Shrewsbury on 12 February 1809, the fifth child of a popular and prosperous local doctor. The grandfather referred to by FitzRoy as 'Dr. Darwin the poet' was Erasmus Darwin (1731–1802), who was not only one of the leading physicians of his generation, but was also a noted poet, inventor, and natural scientist. Among his many achievements, he was one of the first writers to suggest that existing species might have evolved from earlier forms of life. Unlike his grandson, however, Erasmus Darwin was not able to propose a feasible mechanism by which evolution could have occurred.

Charles Darwin commenced his education at Shrewsbury School. Rather fortunately as it turned out, his father removed him from school when he was only 16, and, taking advantage of the fact that Charles's older brother Erasmus was moving from Cambridge to Edinburgh to complete his medical degree, sent Charles along as well, also to study medicine. Although not particularly taken by medicine, Charles became friendly with a number of naturalists who greatly stimulated his interests in natural history, to the extent that in his second year at Edinburgh, he was able to report some original discoveries about marine organisms to the Plinian Society.[15]

After two years at Edinburgh, it was evident that Charles was not going to follow in his father's and grandfather's medical footsteps. Instead, he was sent to Christ's College, Cambridge, with the aim of preparing for the ministry. After three years, he obtained a creditable result in his final exams, ranking tenth in the group of 178 students who passed. More importantly, during the years at Cambridge he became interested in geology, and greatly extended his understanding of natural history. In these extracurricular fields, he gained much knowledge through the close friendships he developed with John Stevens Henslow, the Professor of Botany, and Adam Sedgwick, the Woodwardian Professor of Geology. After passing his final exams, he had to stay in Cambridge for two more terms to satisfy the degree requirements. These he occupied very fully, in effect as a postgraduate student in geology and natural history. He obviously impressed his mentors, for, as FitzRoy's account shows, it was through Henslow that Darwin was offered the post of naturalist on board the *Beagle*.[16]

In addition to FitzRoy and Darwin, there were, as mentioned previously, several other members of the ship's company who are relevant to the story of the *Beagle*'s visit to Australia. One of these was Augustus Earle.

As FitzRoy himself stated:

> *Knowing well that no one actively engaged in the surveying duties on which we were going to be employed, would have time—even if he had ability—to make much use of the pencil, I engaged an artist, Mr. Augustus Earle, to go out in a private capacity; though not without the sanction of the Admiralty, who authorized him also to be victualled.*[17]

Augustus Earle was born on 1 June 1793, in London, the son of an American artist, James Earl.[18] After taking lessons at the Royal Academy, Earle travelled the world as a painter, spending time in the Mediterranean, North America, and Brazil. On 17 February 1824, he left Rio de Janeiro on board the *Duke of Gloucester*, which was bound for the Cape of Good Hope. In the middle of March, the ship called in at the south Atlantic island of Tristan da Cunha. A few days later, much to Earle's consternation, it sailed without him.[19]

After he had spent eight lonely months on the island, with only six other adults for company, his frantic signals to passing ships were finally heeded by the *Admiral Cockburn*, which was bound for Van Diemen's Land. He arrived in Hobart on 18 January 1825. Four months later he was in Sydney, where, among other things, he set up a studio and art gallery at No. 10 George Street. During the next three and a half years, he travelled extensively in Australia and New Zealand, producing many paintings, especially landscapes, as he went. By 1830, he was back in England, and had published his *Views in New South Wales and Van Diemen's Land*. On 28 October 1831, with this wealth of experience behind him, he was appointed draughtsman on the *Beagle*.[20]

Another member of the *Beagle's* company relevant to our story was Syms Covington, who commenced the voyage as 'fiddler & boy to Poop-cabin', but who became Darwin's servant in June or early July, 1833. Covington, who was born around 1813, remained with Darwin throughout the whole voyage of the *Beagle* and for three years after the voyage, providing great assistance in the collection, storage and classification of the many specimens of plants, animals, rocks and fossils they had encountered during the voyage. When he finished working for Darwin, Covington emigrated to Australia and later became Postmaster at Pambula on the south coast of New South Wales, where he remained until his death in 1861.[21]

Augustus Earle, waiting forlornly for a passing ship to rescue him from the island of Tristan da Cunha in the south Atlantic.

Philip Gidley King (Jnr), as a boy of eight in mid-1829, sketched by his father Captain P. P. King, on board the Adventure, during the *Beagle's* first surveying voyage.

By the beginning of December 1831, all preparations for the voyage were complete, and the Beagle was ready to sail from Devonport, which is part of Plymouth harbour. After several false starts due to bad weather, the great adventure finally began. In Darwin's words:

> After having been twice driven back by heavy south-western gales, Her Majesty's ship Beagle, a ten-gun brig, under the command of Captain Fitz Roy, R.N., sailed from Devonport on the 27th of December, 1831. The object of the expedition was to complete the survey of Patagonia and Tierra del Fuego, commenced by Captain King in 1826 to 1830—to survey the shores of Chile, Peru, and of some islands in the Pacific—and to carry a chain of chronometrical measurements round the World.[22]

Once at sea, Darwin soon got to know the officers on board. One of them was Philip Gidley King (Jnr), the son of Captain P. P. King, and the grandson of the third governor of the colony of New South Wales, Philip Gidley King (Snr). King (Jnr) was born at Parramatta on 31 October 1817, and, when five years old, travelled to England with his parents. After spending some time at school, he sailed at the age of eight with his father on the Beagle's first surveying voyage. Following a short period of shore leave after the five-year voyage, he was appointed midshipman on the Beagle's second voyage, and soon struck up a friendship with Darwin. As King recorded in his unpublished autobiography,

> After being a few days at Sea I found a firm friend in the person of Mr Charles Darwin to whom my fancy was to relate my experiences in my former voyage.[23]

In old age, King drew from memory several pictures of the layout of the Beagle's decks. As shown in the illustrations on the following pages, they provide a good impression of the environment in which Darwin and his shipmates were to spend the bulk of the next five years.

It is often not appreciated just how young some of the more important members of the ship's company were at this time; of those who are of particular interest in the following account, FitzRoy was 26, Darwin was 22, Covington was 18, King was 14, and Earle was a relatively old man of 38.

The Beagle's voyage proceeded without undue problems, except that Earle soon became too sick to perform his duties. As FitzRoy reported:

> Mr Earle suffered so much from continual ill health, that he could not remain on board the Beagle after August 1832.[24]

However, all was not lost:

The disappointment caused by losing his services was diminished by meeting Mr Martens at Monte Video, and engaging him to embark with me as my draughtsman.[25]

Conrad Martens had been born in London in 1801, his mother English and his father German. He had studied landscape painting under A. V. Copley Fielding, and was an accomplished artist when he left England for the first time, in May 1833, having been offered a passage on the *Hyacinth*, which was undertaking a three-year cruise to India. Two months after leaving England, the *Hyacinth* was in Rio de Janeiro, where Martens heard that the Captain of the *Beagle*, which was then

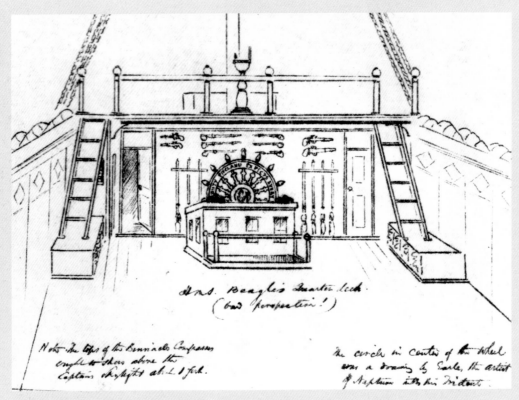

A view of the *Beagle's* quarterdeck, looking towards the poop cabin, drawn from memory by Philip Gidley King (Jnr) in 1891. Darwin worked and slept in the poop cabin, which was only 2.7 metres wide by 1.5 metres deep by 1.8 metres high, and which housed the library of approximately 275 books. John Lort Stokes (mate and assistant surveyor) and King shared the workspace in the poop cabin with Darwin.

at Montevideo, was seeking an artist as replacement for Earle. Martens travelled to Montevideo, offered his services to FitzRoy, and was duly appointed as Earle's replacement.[26]

These negotiations took place during one of Darwin's many inland excursions. FitzRoy reported the good news to Darwin on 4 October 1833, in a high-spirited letter written on board the *Beagle*. The letter says quite a lot about Martens, but is even more interesting for the insight it provides into the relationship between FitzRoy and Darwin:

> *If M^r. P. has written as he intended you have heard of M^r. Martens—Earle's Successor,—a <u>stone pounding artist</u>—who exclaims <u>in his sleep</u> "<u>think</u> of <u>me</u> standing*

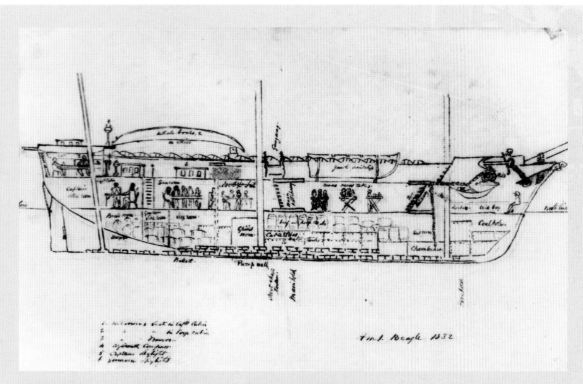

A diagrammatic section of the *Beagle*, drawn from memory by Philip Gidley King (Jnr) in 1891. Some idea of the scale of these drawings can be gained by noting that the *Beagle* was approximately 27 metres long and 7.5 metres wide. After refitting prior to the 1831–36 voyage, her total displacement was approximately 250 tonnes.

A self-portrait of Conrad
Martens at around 33
years of age, when he was
Darwin's shipmate on the
Beagle.

upon a pinnacle of the Andes—or sketching a Fuegian Glacier!!!" By my faith in Bumpology, I am sure you will like him, and like him <u>much</u>—he is—or I am wofully mistaken—a "rara avis in navibus,— Carloque Simillima Darwin".— Don't be jealous now for I only put in the last bit to make the line scan— you know very well your degree is "rarissima" and that <u>your</u> line runs thus— Est avis in navibus Carlos rarissima Darwin.— but you will think I am cracked so seriatim he is a gentlemanlike, well informed man.— his landscapes are <u>really</u> good (compared with London men) though perhaps in <u>figures</u> he cannot equal Earle— He is very industrious— and gentlemanlike in his <u>habits</u>,—(not a <u>small</u> recommendation).[27]

This is a remarkable letter. FitzRoy is excited by his new, enthusiastic artist, and is confident that Darwin will like him, because his head shape (Bumpology = phrenology) is acceptable. So carried away is FitzRoy that he composes a line of Latin verse, declaring that Martens is a rare one on ships ('rara avis in navibus'),[28] and is extremely similar to Charles Darwin ('Carloque Simillima Darwin').[29]

The next sentence of FitzRoy's letter ('Don't be jealous now …') was necessary because he is worried that his use of the superlative form *simillima*, in this context meaning 'extremely similar', will make Darwin jealous: he fears that Darwin may interpret *simillima* as indicating that he is raising Martens to a level of esteem too close to that in which he holds Darwin. FitzRoy realises that the positive form *similis* or the comparative form *similior* would have been less likely to make Darwin jealous, but neither would have fitted his chosen metre.[30]

Determined not to run any risk of offending his good friend, FitzRoy assures Darwin that in fact he is an extremely rare ('rarissima') type of person. This word provides FitzRoy with a solution to his quandary. He composes another line of verse, this time using *rarissima* instead of *simillima*. By stating that Darwin is an extremely rare one on ships ('Est avis in navibus Carlos rarissima Darwin'), FitzRoy achieves his aim of expressing the highest admiration for Darwin.[31]

One is left wondering how many modern ship captains would be able to write such a letter to a friend, and how many modern friends would be able to understand it!

Fearing that Darwin will think him mad ('cracked') for carrying on in such a fantastical manner, FitzRoy dispenses with the frivolities but not the Latin, and presents a list (*seriatim* = one after another) of Martens's good qualities.

A little later in the letter, the Captain explains why he is in such high spirits:

> *I will never write another letter after tea—that green beverage makes one tipsy.*

In light of the strength of feeling that FitzRoy shows for Darwin in this letter, it is sad to think that their relationship broke down completely after the publication of Darwin's theory of evolution, which so upset FitzRoy because of his by-then absolute faith in a literal interpretation of the Bible.

On 13 November 1833 Darwin reported the news of Martens's appointment to his sister Caroline, in a letter written from Montevideo:

> *. . . a Mr Martens, a pupil of C. Fielding & excellent landscape drawer has joined us.— He is a pleasant person, & like all birds of that class.—full up to the mouth with enthusiasm.*[32]

By the middle of the next year, Darwin had some firsthand experience of Martens. In a letter written on 29 July 1834 to another sister, Catherine, from 'a hundred miles South of Valparaiso', he remarked that

> *Our new artist, who joined us at M. Video, is a pleasant sort of person, rather too much of the drawing-master about him; he i<s> very unlike to Earles eccentric character.*[33]

On 8 March 1833, while in the Falkland Islands, FitzRoy had bought at his own expense a ship called the *Unicorn*,[34] in order to complete his surveying tasks more quickly. He renamed this ship the *Adventure*, in memory of the *Beagle*'s sister ship on its previous voyage. The second ship was a great success, but by August 1834, when both ships had reached Valparaiso, FitzRoy was running short of funds. Having received no support from the Admiralty, FitzRoy reluctantly decided that he had no alternative but to sell the *Adventure* and to dismiss its crew. Martens was amongst those who were signed off. Darwin reported this to Caroline in a letter written in Valparaiso on 13 October 1834:

> *You will be sorry to hear, the Schooner, the Adventure is sold; the Captain received no sort of encouragement from the Admiralty & he found the expense <of> so large a vessel so immense he determined at once to <give> her up.— We are now in the same state as when we left England with Wickham for 1st Lieut, which part of the business anyhow is a good job.— we shall all be very badly off for room; & I shall have trouble*

enough with stowing my collections. It is in every point of view a grievous affair in our little world; a sad tumbling down for some of the officers, from 1ˢᵗ. Lieut of the Schooner to the miserable midshipmans birth.—& many similar degradations.— It is necessary also to leave our little painter, Martens, to wander about yᵉ world.— Thank Heavens, however, the Captain positively asserts that this change shall not prolong the voyage.—that in less than 2 years we shall be at New S. Wales.—[35]

Apparently undeterred by this setback, Martens, too, had decided to travel to New South Wales, and took the precaution of obtaining a letter of introduction from FitzRoy. The letter is addressed to Captain P. P. King, who at that time was one of the most influential citizens in the colony. As we have already seen, FitzRoy was well acquainted with King, having served under him during the last two years of the *Beagle*'s previous surveying voyage to South America. In addition, as we have also seen, King was the father of FitzRoy's midshipman, P. G. King. FitzRoy's letter indicates the strong impression made by Martens during his brief sojourn on board the *Beagle*:

H..M.S. Beagle
Valparaiso,
5ᵗʰ. Nov. 1834.

Dear Captain King
The bearer of this letter, Mʳ. Conrad Martens,— has parted from me— I am sorry to say,— because there is no longer room for him on board the Beagle, nor money for him in my pocket.

Had I more money, and more stowage room, I should not think of ending my engagement with him.

He has been nearly a year with us,— and is much liked by my shipmates and myself. He is a quiet, industrious,— good fellow;— and I wish him well.

He thinks of visiting and perhaps settling at Sydney,— therefore I write this letter by way of an introduction, to you. Enclosed is a letter I received about him from Captain Blackwood of the Hyacinth.

You will be able to judge of his abilities, by a glance at his works, far better than by any words of mine.

He has a host of views, of Terra Del, in his sketch book.

His profession is his maintenance.

<signature excised> [36]

Martens certainly did have 'a host of views' of Tierra del Fuego and of other places in southern South America. In fact, he had almost filled four sketchbooks. Some of his *Beagle* sketches were included in the official account of the voyage. But many remained unpublished until 1979 when they appeared in Richard Keynes's *The Beagle Record*.[37]

Armed with this letter and his sketchbooks, Martens embarked on a ship called the *Peruvian* on 3 December 1834, bound for Tahiti, where he stayed for seven weeks. On 4 March 1835, he left Tahiti on board the *Black Warrior*, bound for Sydney. On the way, the ship spent five days at the Bay of Islands, New Zealand. Finally, on 17 April 1835, Martens passed through Sydney Heads and disembarked in Sydney.

Beagle off Mt. Sarmiento, by Conrad Martens. This oil painting is based on one of many sketches of Tierra del Fuego made by Martens when he was artist on the *Beagle*. The sketches were mentioned by Captain FitzRoy in his letter introducing Martens to Captain P. P. King. This particular painting was sold by Martens to King in April 1838.

During the following months he produced many fine sketches, and some paintings, of Sydney and its surrounding districts. He also took several excursions, including one over the Blue Mountains. These sketches and paintings provide a unique set of contemporary illustrations for Darwin's account of his visit to New South Wales.

The *Beagle* remained at its surveying tasks on the west coast of South America for nine months after Martens had sailed for Tahiti. Towards the end of this period, on 3 September 1835, while in Lima, Peru, Darwin wrote a letter to his sister Susan. In it, he showed that he was quite looking forward to visiting Sydney:

> *Two men of war have lately arrived from Rio, but they brought no letters for the Beagle; so that the Admiral is forwarding them on to Sydney.— We all on board are looking forward to Sydney, as to a little England: it really will be very interesting to see the colony which must be the Empress of the South.— Capt King has a large farm, 200 miles in the interior.— I shall certainly take horse & start—I am afraid however there are not Gauchos, who understand the real art of travelling.*[38]

As we shall see, Darwin did take a trip inland from Sydney, but only as far as Bathurst. The Captain King mentioned by Darwin is once again Captain P. P. King. Although King had financial interests in large areas of land in the colony, he actually lived on a farm called Dunheved near Penrith, which is only 30 miles (50 kilometres) west of Sydney.

Having completed its surveying tasks in South America, the *Beagle* headed west-wards into the Pacific Ocean on 8 September 1835, following its next set of instructions, which were to survey the Galapagos Islands (where Darwin made important obser-vations concerning the variation in species between islands), and then, as we have seen, to take chronometer readings at various islands across the Pacific until reaching Sydney, where her chronometers were to be checked at Parramatta Observatory.

So it was that the *Beagle* sailed across the Pacific Ocean, bound for Sydney. On the way, she stopped in New Zealand for ten days, anchoring in the Bay of Islands, which at that time was the only European settlement in that country. Three days before departing from New Zealand, Darwin wrote a letter to his sister Caroline. It provides a good indication of how he felt at that time, and of how much he was looking forward to visiting Sydney:

> *Bay of Islands.— New Zealand.*
> *Decemb 27ᵗʰ. 1835.—*
>
> *My dear Caroline,*
> *My last letter was written from the Galapagos, since which time I have had no opportunity of sending another. A Whaling Ship is now going direct to London*

& I gladly take the chance of a fine rainy Sunday evening of telling you how we are getting on.— You will see we have passed the Meridian of the Antipodes & are now on the right side of the world. For the last year, I have been wishing to return & have uttered my wishes in no gentle murmurs; But now I feel inclined to keep up one steady deep growl from morning to night.— I count & recount every stage in the journey homewards & an hour lost is reckoned of more consequence, than a week formerly. There is no more Geology, but plenty of sea-sickness; hitherto the pleasures & pains have balanced each other; of the latter there is yet an abundance, but the pleasures have all moved forwards & have reached Shrewsbury some eight months before I shall.—

If I can grumble in this style, now that I am sitting, after a very comfortable dinner of fresh pork & potatoes, quietly in my cabin, think how aimiable I must be when the Ship in a gloomy day is pitching her bows against a head Sea. Think, & pity me.— But everything is tolerable, when I recollect that this day eight months I probably shall be sitting by your fireside.— . . .

338   APPENDIX.

Bay of Islands in New Zealand to Sydney.

Fifteen Chronometers. Nineteen Days.

	H. M.	S.	H. M.	S.		H. M.	S.	H. M.	S.
A	1 31	33,50	......	33,50	N	1 31	32,52	......	32,52
B	......	27,63			O	......	29,29	......	29,29
C	......	33,64	......	33,64	R	......	30,69	......	30,69
D	......	31,84	......	31,84	S	......	30,17	......	30,17
G	......	27,41			W	......	28,52	......	28,52
H	......	23,94			X	......	26,82		
K	......	44,60			Z	......	32,36	......	32,36
L	......	32,09	......	32,09					
					Mean ...		31,00	......	31,46

Preferred............ 1h. 31m. 31,5s.

Places of observation :

New Zealand, as before.

Sydney, Fort Macquarrie.

From Macquarrie Fort to the Observatory at Paramatta, by three Chronometers,* 0h. 00m. 52,0s. (Paramatta west of Fort).

*These three Chronometers were carried by water to and from the Observatory on the same day.

Readings recorded on the *Beagle*'s chronometers, used to estimate the difference in longitude between Bay of Islands in New Zealand, and Fort Macquarie (now the Opera House) in Sydney. The last two lines and the footnote indicate that three chronometers were taken to the observatory at Parramatta and returned to the *Beagle* in the one day. The words 'carried by water' confirm that the trip was made via the Parramatta River.

. . . I am looking forward with more pleasure to seeing Sydney, than to any other part of the voyage.— our stay there will be very short, only a fortnight; I hope however to be able to take a ride some way into the country.— From Sydney, we proceed to King George's sound, & so on as formerly planned. . . .

Your's
C. Darwin[39]

The scene is now set for Charles Darwin's visit to Australia.

Arrival in Sydney and a trip across the Blue Mountains

As the lighthouse on South Head came into view at ten minutes past eleven on the evening of 11 January 1836, the officers and crew of the *Beagle* caught their first glimpse of the entrance to Port Jackson.[1] In the daylight of the following morning, the ship entered Sydney Harbour.

Jan 12ᵗʰ Early in the morning, a light air carried us towards the entrance of Port Jackson: instead of beholding a verdant country, a straight line of white cliffs brought to our minds the shores of Patagonia. A solitary light-house, built of white stone, alone told us, we were near to a great & populous city.

(DIARY)

This imposing lighthouse was Francis Greenway's Macquarie Light. Built in 1818, it continued in operation until 1883, when, owing to structural problems, it was replaced by a replica, which still stands today.

This scene is part of a sketch drawn by Conrad Martens as he approached the Heads for the first time on 17 April 1835. It is Martens's first Australian sketch, and it shows the same view that Darwin saw on the morning of 12 January 1836, as the *Beagle* approached Australia for the first time.

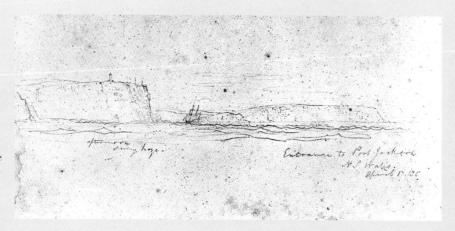

'A solitary light-house, built of white stone, alone told us, we were near to a great & populous city.'

Darwin was referring to Francis Greenway's famous Macquarie Lighthouse on South Head, seen here in a sketch drawn by Conrad Martens on 6 December 1835, just five weeks before it attracted Darwin's attention.

Darwin's Diary continues:

Having entered the harbor, it was a fine, spacious, appearance; but the level country, showing on the cliff-formed shores, bare & horizontal strata of sandstone, was covered by woods of thin scrubby trees that bespoke useless sterility.— Approaching further onwards, patches of the country improved; everywhere very beautiful Villas & nice cottages were scattered along the beaches. Large white stone houses, two & three stories high, & Wind-mills standing along the edge of a bank, pointed out to us the Capital of Australian civilization which had not yet come into view.—

'Having entered the harbor, it was a fine, spacious, appearance.'

This Martens painting shows a view of Sydney Harbour, looking westward from near the Macquarie Lighthouse on South Head. It was painted in the same year that Darwin visited Sydney, and is based on a sketch made by Martens four days after the *Beagle* departed for Hobart. The peninsular in the right foreground is Nielson Park leading down to Bottle and Glass Point. Beyond that is Bradley's Head.

Among the houses that Darwin saw were Henrietta Villa, on a headland called Eliza Point (now Point Piper), and the many dwellings recently erected by the colony's leading citizens on Woolloomooloo Hill, which had become the social centre of Sydney.[2]

At 1.30 p.m., with heavy clouds and occasional lightning overhead, the *Beagle* shortened all sails and came to in Sydney Cove:

> *At last we anchored within Sydney Cove; we found the little basin containing many large ships & surrounded by Warehouses. on one point stood an insignificant little fort.—*

Sydney Cove as viewed from Kirribilli Point at the time of Darwin's visit. This Conrad Martens watercolour is based on a sketch made only four days before the *Beagle* arrived. From left to right, the buildings identified from Martens's sketch are a Bathing House (on the water's edge), Roman Catholic Chapel (on the horizon; the first St Mary's), Hyde Park Barracks (now the Hyde Park Barracks Museum), Part of Council Chambers, Government Stables (now the Conservatorium of Music), Old Mill, Church (St James), Macquarie Fort (the site of the Opera House), Sydney Barracks (on the horizon behind the Cove), and Government Warehouses. Many of these buildings are shown on the map on p. 27.

In a letter to his sister Susan, Darwin described his feelings at that time:

> *On coming to an Anchor I was full of eager expectation; but a damp was soon thrown over the whole scene by the news there was not a single letter for the Beagle.— None of you at home, can imagine what a grief this is. There is no help for it: We did not formerly expect to have arrived here so soon, & so farewell letters.— The same fate will follow us to the C. of Good Hope; & probably when we reach England, I shall not have received a letter dated within the last 18 months. And now that I have told my pitiable story, I feel much inclined to sit down & have a good cry.*[3]

ARRIVALS.

From Lima via New Zealand, on Tuesday last having left the former place on the 6th September and the latter on the 30th ultimo, H. M. Ship *Beagle*, Captain Fitzroy. This vessel has been employed surveying the Coast of South America.

From Hobart Town, same day, whence she sailed the 1st instant, the barque *Leda*, Captain M'Load with sundries.

From Norfolk Island, same day, whence she sailed the 30th ultimo, the Government schooner *Isabella*, Captain Boyle with sundries. Passengers, Mr. Plunkett, Lady, and servants, Mr. Fisher and servant. Rev. Messrs Ullathorne and Styles, Captain Best, Lieutenants Briggs and Gibson, one Ticket of Leave man, and the Executioner.

The *Beagle*'s arrival was not an important event in the life of the colony, and it received only brief mention in the local press. Shown here is an extract from the 14 January 1836 edition of the *Sydney Gazette*.

The *Beagle* at anchor in Sydney Harbour. This sketch was drawn by Martens when the *Beagle* returned to Sydney on its third surveying voyage, in 1839.

Although there was no news from England, there was a local letter of some importance, addressed to Darwin's friend P. G. King. As we have already seen, King had left the colony when only five years old, and had spent nine of the thirteen intervening years at sea. Now a young man of eighteen, he was finally returning to his birthplace to be reunited with his parents. Knowing that they lived almost 30 miles (50 kilometres) west of Sydney on a farm called Dunheved, and realising that they had no way of knowing when the *Beagle* would arrive, King would not have expected his parents to come to Sydney to greet him. They did, however, send a letter to the *Beagle*, telling King to go ashore and call on his former shipmate Conrad Martens.[4] As King explains in some autobiographical notes:

> *Mr Conrad Martens had been our Artist in the Beagle but he left us to go across the Pacific to Sydney with an introduction to my Father — and he established his studio in Bridge Street, whither I went on landing and found instructions from my Father how to get to Dunheved.*[5]

Following his father's instructions, King contacted a Mr J. B. Jones, who drove him to Vineyard, a property on the northern bank of the Parramatta River, about halfway between Sydney and Dunheved. Waiting to greet him there were his aunt and uncle, Anna Maria and Hannibal Macarthur, and his fourteen-year-old brother Essington, eager to escort him home to Dunheved:

> *J. B. Jones drove me to Vineyard and after inspection by my Uncle Hannibal my Aunt and their four daughters, I was driven in his gig to Dunheved. Essington accompanying me on horseback and galloping on to announce my approach — ! and a happy greeting I got — after my long absence.*[6]
>
> *. . . I met [my father] on a road up from Dunheved at the [slip rail] with my Uncle Copland who had walked up from the house to meet me on the 12th January, 1836. I thought him at first much aged but this impression wore off after a day or two. My mother had waited for me just outside the house. I had not seen her since 25 May, 1826. My feelings at finding myself with a Mother after such a Separation were as strange as the situation was novel.*[7]

King's mother was overjoyed at being reunited with her eldest son. On 17 February 1836 she recorded in her diary:

> *I cannot but record the feelings I have lately experienced, my dear Philip Gidley after an absence of 10 years is restored to me in health and strength, and apparently very amiable.*[8]

Bridge Street in Sydney, at the time of the *Beagle*'s visit. Martens established his first studio here in 1835, and both Darwin and P. G. King (Jnr) visited him here soon after the *Beagle* arrived in Sydney Cove. This Martens watercolour is based on a sketch drawn just a few weeks earlier, in December 1835.

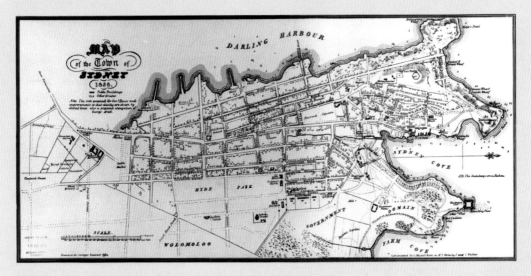

A map of the town of Sydney during the time of Darwin's visit.

In the meantime, in Sydney, Darwin recovered from his disappointment at the lack of mail, and went ashore. In the words of his Diary:

In the evening I walked through the town & returned full of admiration at the whole scene.— It is a most magnificent testimony of the power of the British nation: here, in a less promising country, scores of years have effected many times over, more than centuries in South America.— My first feeling was to congratulate myself, that I was born an Englishman:— Upon seeing more of the town on other days, perhaps it fell a little in my estimation; but yet it is a good town; the streets are regular, broard, clean & kept in excellent order; the houses are of a good size & the Shops excellent.— It may be compared with great accuracy to the large suburbs, which stretch out from London & a few other great towns:— But not even near London or Birmingham, is there an aspect of such rapid growth; the number of large houses just finished & others building is truly surprising; & with this, every one complains of the high rents & difficulty in procuring a house.—

In the streets Gigs, Phaetons & Carriages with livery Servants are driving about; of the latter vehicles, many are as neat as those in London.— Coming from S. America, where in the towns, every man of property is known, no one thing surprised me more, than not readily being able to ascertain to whom this or that Carriage belonged.— Many of the older Residents say that formerly they knew every face in the Colony, but now that in a morning's ride, it is a chance if they know one.—

Sydney has a population of 23 thousand & is as I have said rapidly increasing; it must contain much wealth; it appears a man of Business can hardly fail to make a large fortune; I saw on all sides large houses, one built by the profits from Steam-Vessels, another from building, & so on. A convict Auctioneer is said intends to return home & will take with him 100,000 £.— Another Convict, who is always driving about in his carriage, has an income so large, that nobody ventures to guess at it.— But the two crowning facts are; that the public revenue has increased 60,000 £ during this last year & that less then an acre of land within the town sold for 8000 £.—

The ex-convict auctioneer was probably Abraham Polack, who had acquired his fortune in less than ten years. The other ex-convict mentioned by Darwin was Samuel Terry. Sentenced to seven years' transportation in 1800 for stealing four hundred pairs of stockings, he began to amass his fortune even before his sentence had expired. After release, he became a hotel licensee, shopkeeper, and speculator, eventually owning property throughout the colony. He was 'always driving about in his carriage' because a stroke two years earlier had left him disabled. When he died two years after Darwin's visit, he was called the Rothschild of Botany Bay, and was reckoned by many to be the wealthiest man in the colony, leaving an estate worth around £500 000.[9]

As well as being struck by the prosperity of the settlement, Darwin was impressed by the public gardens:

> *There is one advantage which the town enjoys in the number of pleasant walks in the Botanic Garden & Government domain; there are no fine trees, but the walks wind about the Shrubberies & are to me infinitely more pleasing than the formal Alamedas of S. America.—*

As Darwin had mentioned earlier in letters[10] to his sisters, he intended to travel inland from Sydney, to see 'the interior' of Australia. After only three days in the colony, he had arranged an expedition and was ready to start. In the Diary, Darwin recorded:

> *I hired a man & two horses to take me to Bathurst, a village about 120 miles [190 kilometres] in the interior & the centre of a great Pastoral district; by this means I hoped to get a general idea of the country.—*

This view of the Domain was sketched by Martens in September 1835.

When describing this excursion in a letter to his sister Susan, Darwin explained:

> My object was partly for Geology, but chiefly to get an idea of the state of the colony, & see the country. Large towns, all over the world are nearly similar, & it is only by such excursions that the characteristic features can be perceived.[11]

On some of his earlier expeditions during the *Beagle*'s port calls in South America and in the Pacific, Darwin had taken his servant, Syms Covington, as his companion. On this occasion, however, Covington stayed in Sydney, possibly because Darwin felt that he needed someone familiar with the route to Bathurst. Unfortunately, Darwin makes no mention of the identity of his companion.

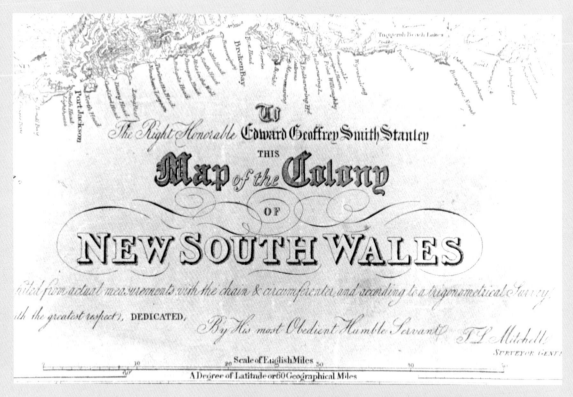

A section of Major Mitchell's 1834 map of the colony, which was the standard map in use at the time of Darwin's visit. Its east-west orientation makes it somewhat difficult for modern readers to interpret. Mitchell presented a copy of this map to FitzRoy just before the *Beagle* departed (see p. 118).

It is almost certain that Darwin would have carried with him an up-to-date map of the colony, most likely the 1834 map compiled by the explorer Major T. L. Mitchell, who had been Surveyor-General since 1828. Darwin later corresponded with Mitchell on questions concerning the geology of the Blue Mountains, and other related topics.

In addition to a map, it is most likely that he also took a copy of the *1835 New South Wales Calendar and General Post Office Directory*, which contained a detailed itinerary of the trip to Bathurst. Providing a graphic description of the route taken by Darwin, this itinerary is reproduced in the present book together with modern maps, so that readers can compare Darwin's journey with a present-day trip to Bathurst.[12]

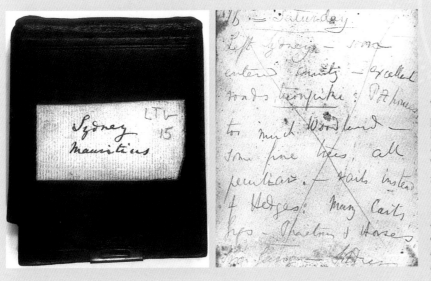

The front cover and first page of the pocket note-book carried by Darwin during his trip to Bathurst. The text on the first page reads:

16th = Saturday
Left Sydney — soon
entered country — excellent
roads, turnpike: Pot houses
too much Woodland —
some fine trees, all
peculiar. — rails instead
of Hedges: many Carts
Gigs — Phaetons & Horses
[illeg] Garrison soldiers

'Pot houses' was a term for pubs. The large cross on the page was presumably drawn by Darwin when he had finished making use of these notes in writing up the relevant passage in his Diary.

Whenever Darwin went on an inland expedition, he carried a field notebook in which he jotted down words and phrases that formed the basis of the Diary that he kept throughout the *Beagle*'s voyage. The notebook fitted conveniently into his pocket, and was always readily available for recording anything of interest. On the inside of the cover, a label declared that this was not just any ordinary notebook: it was a 'Velvet Paper Memorandum Book, so prepared as effectually to secure the writing from erasure; with a Metallic Pencil the point of which is not liable to break'.

By the time the *Beagle* had reached Sydney, Darwin had filled at least fourteen of these books. For the trip to Bathurst he started a new one, which he later called the Sydney/Mauritius notebook. Along with the other field notebooks, it is now preserved in Down House, formerly Darwin's home, in the village of Downe in Kent.

The entries in this notebook are Darwin's first comments on Australia. Written on the impulse of the moment, and often in a great hurry, the words are not easy to decipher, and some are illegible. Despite these difficulties, the Australian section of the notebook has been transcribed, and it provides a record of the thoughts that were passing through his mind during the trip to Bathurst. Extracts are quoted as supplements to the Diary throughout the remainder of this and the next chapter.

When Darwin and his companion commenced their journey on the fourth day after the *Beagle*'s arrival in the colony, it was a fine sunny morning in Sydney, with light winds blowing from the south.[13]

The notebook commences:

> *16th = Saturday*
> *Left Sydney — soon entered country — excellent roads, turnpike.*

After a few hours, the travellers had reached Parramatta where they

> *Lunched nice little Public House.*

After lunch, they continued their

> *Ride on to Emu ferry on the Nepean: a broard river still as a pool, small body of running water.*

In his Diary, he expands on these observations:

> *In the morning of the 16th. I set out on my excursion; the first stage took us through Paramatta a small country town, but second to Sydney in Australia.— The roads were excellent & made on the Macadam principle; The whinstone with which they are made is brought from the distance of several miles. There are turnpikes.— The road appeared much frequented by all sorts of Vehicles.— I met two Stage Coaches.—*

ITINERARY;

COMPRISING THE ROADS THROUGHOUT

NEW SOUTH WALES.

THE Traveller, on setting out for any part of the Colony, except for South Head or Botany Bay, must leave Sydney by George-street, which extends nearly a mile.

⅓ Turnpike, 1⅓.

1½ A little beyond, on the left, is the road leading to Botany Bay, where Mr. Simeon Lord has a mill and woollen manufactory. Near where this road to Botany Bay turns off, the line so long talked of, as the proper direction for the Great Road from Sydney—*the trunk to which all other Roads are but branches*—would proceed along the fine flats, *still open*, to the south-east of Mr. Shepherd's nursery. A plan for the regular extension of Sydney to its new limits in that direction being (as it is reported) in contemplation, the traveller must picture to himself how different the scene may become on that side in a few years. ½.

On the same side of the way, further on, is the Spinning-wheel public house, and what is called the "Soldier's Garden," belonging to Government. The Brisbane Distillery (Mr. R. Cooper) is an extensive pile of building, where the business is now carried on, and an excellent spirit produced, very similar to London gin.

Opposite the Brisbane Distillery, on the right, is the gate leading to Ultimo House, belonging to Dr. Harris (the present residence of J. E. Manning, Esq.), who has some hundred acres of land here, a portion of which has recently been sold in building allotments. Further on, is land belonging to the late John Macarthur, Esq., marked by an old windmill, which is a very picturesque object from various parts about Sydney. On the margin of Darling Harbour stands Newstead, a commodious house and premises recently

built by the late George Bunn, Esq.,—a gentleman whose early death was alike a loss to his amiable family, his numerous friends, and the country at large, and whose public spirit and private worth could not be too publicly or permanently recorded; alas, that the words of Ossian, quoted by Byron respecting another Newstead, should apply also to the builder of this!

"Why dost thou build the hall? Son of the winged days! Thou lookest from thy tower to-day; yet a few years, and the blast of the desert comes; it howls in thy empty court."

1¾ Bridge across Black Wattle Swamp Creek, ¼.

The road leading to Petersham and Cook's River on the left; a little further along this road are the gardens and nursery of Mr. Shepherd, a very spirited practical horticulturist, a high-priest of Flora and Pomona, on whose success and encouragement depends the quality of the contents of the Cornucopia of Australia; there may be seen, in embryo, the future olive-grounds and vineyards which are to be spread over the valleys of Australia, with many rare fruits, flowers, and plants from India and China; nor will there be any want of roses for fair fingers to twine, the most beautiful varieties of this prince of flowers are to be seen there, produced by Mr. Shepherd's tasteful cultivation. This road passes by the back of the Race Course, and is fair travelling for a gig as far as the ford at Cook's River, across which are Tempé, the residence of A. B. Spark, Esq., and other country seats. The bush roads from this ford open into the country which lies between Cook's and George's Rivers.

2 On the right, the Glebe Road, so named because it leads through lands sold about three years ago by the Church; several good houses have been erected lately, and gardens formed. At the junction of the roads is the Archdeaconry, a portion of land where it was contemplated to erect a residence for the Archdeacon of New South Wales. On the left are paddocks and fields belonging to Grose Farm, a Government establishment.

Bridge across the Orphan School Creek, which is here the western boundary of the Glebe.

3 Adjoining the boundary of Grose Farm on the left, is the old Sydney race course.

Bridge across Johnson's Creek.

On the left, Bagnigge Wells public house, kept by J. Sappe.

3½ On the left is Annandale, the residence of Robert Johnstone, Esq., with an avenue of Norfolk pines leading to the house; it is one of the most complete farms in the neighbourhood of Sydney, ½.

3⅔ Bridge.

4¼ On the right, road to Birch Grove, a pleasant seat on the Paramatta River.

4½ On the right is Elswick, the seat of James Norton, Esq. The house was built, and the grounds laid out, by Mr. James Foster, a place which, taken altogether, is laid out in a style more truly *English* than any other in the Colony.

On the left is the road to Petersham, the country residence of the lamented Dr. Wardell, whose recent melancholy death is a subject too painful to notice further.

4¾ { Cherry Gardens public house on the left.
Cheshire Cheese ditto on the right, ¼.

[These houses have Tea Gardens, &c.; some *fêtes a la Vauxhall* were once given there.]

4⅛ Long Cove Bridge.

5¼ { Public house (Speed the Plough) to the right, recently established.
Great South Road through Liverpool, to the left. *(See Great South Road.)* p 109.

6 To the right, a little further on, is Dobroyd, the residence of Doctor Ramsay.

6¾ Iron Cove Bridge to the right, where the new North Road, across the Paramatta River by a punt, separates *(see the new North Road)*, which would shorten the distance many miles to travellers northward, were this road opened throughout, 1.

The Ship Inn, or "Half-way House," to the left.

Ashfield Park to the left, the residence of the late Mr. Joseph Underwood.

7 An opening to the right, which has been made in the bush, shews the waters of Hen and Chicken Bay, an inlet of the Paramatta River.

The Fox and Grapes Inn to the left.

7½ Bridge across creek, entering to the left, Hen and Chicken Bay.

A little further, lodge and gate to the right, leading to Government farm, Longbottom.

7¾ Gate of Burwood on the left.

8¼ The road to Concord and Lovedale on the right, on one of the arms of the Paramatta River, not far from Kissing Point, distant two miles, ½.

8¾ Powell's Bridge.

Coach and Horses public-house.

9⅓ Bush road to the left, leading through the Glebe land, Liverpool Road, which it enters near the nine mile stone.

Another bridge.

The cleared ground of Home Bush on the right, formerly the residence of D'Arcy Wentworth, Esq. being a mile and a half, which is the greatest extent of land cleared and stumped in the neighbourhood of Sydney. The hills seen beyond are the Pennant Hills, across the Paramatta River, where there are numerous small farms, formerly known as the "Eastern Farms."

11 Small bridge.

On the left Traveller's Inn.

Bush tract (left) to the Dog Trap Road.

11¼ Haslem's Bridge, on Hacking's Creek, which is the Western boundary of Home Bush. At this bridge the proposed new road from Sydney would either join the old road, or cross it, by which it would enter the lower part of Paramatta in a more direct line from Sydney, 3.

Lodge and avenue (right) to Newington, the residence of John Blaxland, Esq., M. C. where extensive salt works are carried on. Mr. Blaxland has lately built a new house, and made several other improvements on this estate.

12¾ Bush Road (left) joining that above-mentioned to the Dog Trap Road.

Small bridge.

12⅞ Duck River Bridge.

13¼ Duck Creek Bridge.

Public-house to the left called Cottage in the Grove. Gate on the right, leading to the residence of the late John Macarthur, Esq.

Turnpike Gate.

14 On the left, Dog Trap Road leading to Liverpool.

Turn to the right over Becket's Bridge, from whence we ascend the hill. On the left is a public-house, and entrance to the Paramatta race course. The first view of the Town. The Female Orphan School, Barracks, &c. are seen.

14½ Bridge over Clay Creek, ½.

'The first stage took us through Parramatta ...'

The New South Wales Calendar and General Post Office Directory was published each year, and contained a detailed itinerary for each of the main roads in the colony. It is very likely that during his trip to Bathurst, Darwin used the 1835 itinerary shown here. His route is shown on the modern map on page 34.

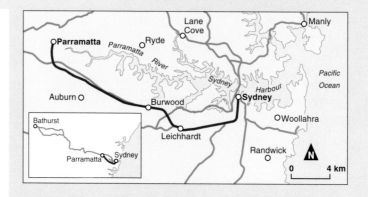

134   ITINERARY.

Great Western Road.   *(See that road.)*

15 The fifteen mile brings the traveller into the middle of
Paramatta, passing the Church, a curious structure, with
a double steeple ; there is also a Government-house, the
occasional retreat of the Governor ; attached to it is an
Observatory, at present used by Mr. Dunlop, on whose
vigilance science may rely with confidence respecting the
Astronomical news of the Southern Hemisphere. The Female Factory is also at Paramatta. The Female Orphan
School, Barracks for soldiers and prisoners. There is a
resident Police Magistrate in this Town. Also some excellent Inns, Walker's "Red Cow" has been long celebrated ;
and the "Woolpack," Nash, is also a commodious inn.
Paramatta is the head quarters of a Regiment ; and Courts
of Request and Quarter Sessions are periodically held here.

Along this road the bridges are on salt water creeks, from
that arm of the sea named "the Paramatta River,"
which forms the harbour of Sydney or Port Jackson. A
new road to Paramatta has been proposed by the Surveyor
General, which would head all these creek and avoid the
steep pulls, as well as the bridges on these creeks, and
thus save the expense of keeping the latter (about twelve
in number) in repair, while the road to Paramatta would
be shortened one and a half mile ; that to Liverpool three
miles. One of the old bridges fell during the last year,
and it is very probable that every year will bring about
other similar interruptions of the thoroughfare along this
most important, but ill-planned road.

A view of Parramatta as it appeared in the year following Darwin's
visit. This Martens painting is based on a sketch drawn on 25
September 1837, looking west from the grounds of Hannibal
Macarthur's property Vineyard, on the banks of the Parramatta River,
about a mile downstream from the town.

In all these respects, there was a most close resemblance to England; perhaps the
number of Pot-houses was here in excess.

While travelling the 24 kilometres from Sydney to Parramatta, Darwin
passed seventeen public houses ('Pot-houses') and one distillery. In the previous few
days, he had seen some of the 198 licensed premises that were then operating in
the town of Sydney, and he was shortly to have lunch in one of the seventeen public
houses in Parramatta.[14] He may not have been the least surprised to learn that the
largest single item in the estimated government revenue for 1836 was £105 000 from
customs duty on spirits, accounting for more than one-half of the total estimated
government income of £199 300; and that, when income from licences for public
houses and from duty on locally produced alcohol was considered together with the
customs duty, nearly 60 per cent of the total government income was derived from
the colonists' consumption of alcohol.[15]

Even if the number of public houses seemed somewhat excessive, their existence
was at least something that reminded him of England. In contrast, the chain gangs
of convicts were completely foreign to the eyes of an Englishman:

> *The most novel & not very pleasing object are the Iron gangs; or parties of Convicts,*
> *who have committed some trifling offence in this country;[16] they are dressed in yellow*
> *& grey clothes, & working in Irons on the roads; they are guarded by sentrys with*
> *loaded arms.— I believe one great means of the early prosperity of these Colonies*
> *is Government thus being able to send large partys at once to make good means of*
> *communication nearer the Settlers.*
>
> *I slept at night at a very comfortable Inn at Emu ferry, which is 35 miles*
> *[55 kilometres] from Sydney & not far from the ascent of the Blue Mountains.—*

(DIARY)

Although this inn is not identified in either the notebook or the Diary, it is
almost certainly the Governor Bourke Inn, which was located where the Log Cabin
Hotel now stands in Memorial Avenue on the eastern bank of the Nepean River, just
south of the present Victoria Bridge at Penrith. According to the itinerary in the
1835 *Calendar*, the Governor Bourke Inn offered 'excellent accommodation'.

Having arrived at Emu Ferry, Darwin describes at length the landscape
through which he has passed that day:

> *This line of road is of course the best & longest time inhabited in the Colony:— The*
> *whole land is enclosed with high railings; for the Farmers not having been able to*

rear hedges.— There are many substantial houses & cottages scattered about; but although considerable pieces of the land [are] under cultivation, the greater part yet remains as when first discovered.— Making allowances for the cleared parts, the country here precisely resembles, all that which I saw during the ten succeding days. — The extreme uniformity in the character of the Vegetation, is the most remarkable feature in the landscape of all parts of New S. Wales.— Every where we have an open woodland, the ground being partially covered with a most thin pasture.

The trees nearly all belong to one peculiar family; the foliage is scanty & of a rather peculiar light green tint; it is not periodically shed; the surface of the leaves

'The most novel & not very pleasing object are the Iron gangs; or parties of Convicts'

This is Augustus Earle's 1830 lithograph of a government jail gang in Sydney.

ITINERARY. 147

GREAT WESTERN ROAD, FROM SYDNEY TO EMU FORD.

$14\frac{1}{2}$ On the right, leave the Parramatta Road.

On the left is a Government paddock of sixty acres for the extension of the town of Paramatta, $14\frac{1}{2}$.

$14\frac{3}{4}$ On the right, the residence of the Reverend Mr. Marsden, $\frac{1}{4}$.

15 On the right, Pitt Row, bounding part of the Government Domain, $\frac{3}{4}$.

On the left, road leading to the Church and School Estate, on the Dog Trap Road. A little further on is the turnpike, and on the right is an entrance into the Government Domain, and the signal station which communicates with Sydney by that of Bedlam.

$15\frac{1}{4}$ On the left, Kenyon's Road, leading across Kenyon's Bridge, and by a path through the Orphan School ground, into the Old Cowpasture Road, which it enters at $25\frac{1}{2}$ miles. The distance is about nine mile miles, $\frac{1}{4}$.

$16\frac{1}{4}$ Bridge across Toongabbee Creek, which forms the Western boundary of the Domain of Paramatta, 1.

On the left, Prospect Old Road, a continuation of Cowpasture Old Road.

17 Bridge across Toongabbee Creek. The road continues through the land of Captain Wentworth, 63d. Regiment, which is cleared of timber, and partly stumped, $\frac{3}{4}$.

18 Bridge over a small creek, 1.

19 On the right, the Richmond New Road. The Old Richmond Road also turns off here, and is now named the Seven Hills Road; it joins the Toongabbee Old Windsor Road at about four miles from where it leaves this road, 1.

20 On the right, a public house, kept by Peisley, named The Fox under the Hill, 1.

On the left, gate leading to Veteran Hall, the residence of William Lawson, Esq. In this neighbourhood, named Prospect Hill, the land consists of one of the richest mineral soils in the Colony. Mr. Lawson has cultivated extensively for many years, and his establishment is very complete. He keeps his best horses on this estate, with other cattle of the most improved breeds.

21 On the right, Flushcombe, the property of Mr. J. Connor, 1.

22 Bridge over a small run of water, 1.

On the right, road which joins the New Richmond Road at $24\frac{3}{4}$ miles. Distance about $3\frac{1}{2}$ miles. This road leads to Bungarabbee House, the property of Mr. C. Smith, leading also to the residence of Robert Crawford, Esq., called Hill End.

148 ITINERARY.

$23\frac{1}{2}$ Bridge across Eastern Creek, $1\frac{1}{2}$.

$23\frac{3}{4}$ On the left, public house, called the Corporation, kept by Mr. Lumpy Dean, $\frac{1}{4}$, and on the same side is Wall Grove, the property of Mr. Charles Roberts.

24 On the right, the clean and comfortable little Inn, appropriately named The Traveller's Rest, kept by Mr. Nixon, whose polite attention to all his customers, entitle him to the patronage of the public.

24 On the right, Rooty Hill, where 8,000 acres have been granted to the Church; most of it is cleared land, having been formerly a Government grazing establishment, $\frac{1}{4}$. The house, and 1,750 acres are now let on lease.

On the left, Minchinbury, at present the residence of Mr. Henry Howey. A sort of grey calcareous rock, resembling marble, has been found on this estate, in quarrying for the roads; on the same side is Mount Philo, the property of the late J. T. Campbell, Esq., now belonging to Mr. W. Hayes.

$26\frac{1}{2}$ On the right, Mount Druitt, the residence of Major Druitt, a handsome looking mansion, $2\frac{1}{2}$.

27 Bridge across Roper's Creek, $\frac{1}{2}$.

On the right, road leading to Dunheved, the seat of Captain P. P. King, R.N.; and a mile further on, Werrington, the seat of C. Lethbridge, Esq.

29 On the right, cross road leading to the farms of Doctor Harris, and his residence, Shane's Park, on the banks of the South Creek, 2.

On the left, a road leading to Erskine Park, the property of the late Colonel Erskine.

$29\frac{1}{4}$ Bridge across South Creek. This creek enters the Hawkesbury at Windsor, i~~t is navigable as high as this, the fallen timber, however, frequently obstructs the navigation.~~

32 Cross the road from Camden to Richmond, at $24\frac{1}{2}$ miles. To the right, it leads to Castlereagh, to Regent Ville, and to the vale of Mulgoa, in the contrary direction, 1.

On the left is Tindale's, ~~a very good inn.~~ on the r. Mr. Machenry

33 ~~On the right,~~ Penrith Post Office and Court House, 1.
~~A little further on is~~ the Governor Bourke Inn, kept by Mr. Wilson, where there is excellent accommodation.

$34\frac{3}{4}$ Emu Ford, a ferry on the Nepean. On the opposite side are Emu Plains, where the Government had a large Agricultural Establishment, and where the Township has been laid out on a portion of rising ground behind the Government farm, across which the new pass, to avoid Lapstone Hill, has been lately formed. About two miles up the river, is Edinglassie, the seat of Chief Justice Forbes. A fine reach of the river immediately above this residence, is capable of floating ~~a dozen~~ men of war, and is navigable for many miles upwards, $1\frac{1}{2}$. (see p 184)

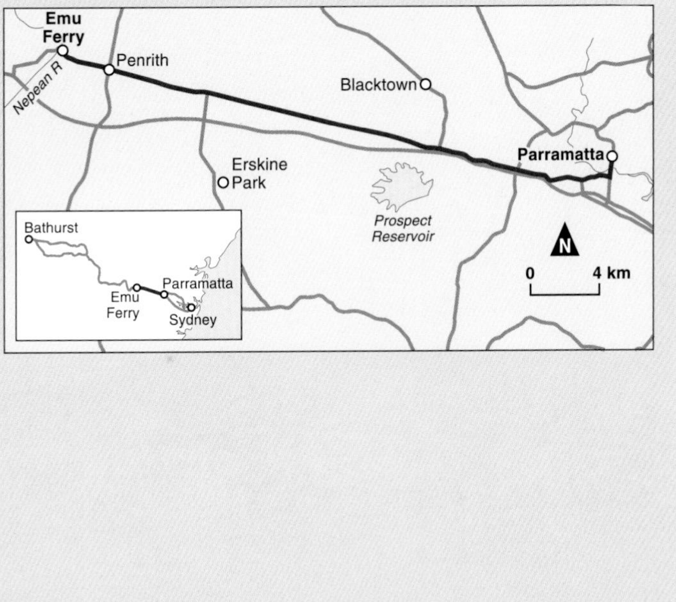

The 1835 *Calendar* itinerary and a modern map of the second part of Darwin's journey on 16 January, from Parramatta to Emu Ford. The 'very comfortable inn at Emu ferry', in which Darwin stayed the night, was the Governor Bourke Inn.

are placed in a vertical, instead of as in Europe a nearly horizontal position; This fact & their scantiness makes the woods light & shadowless; although under the scorching sun of the summer, this is a loss of comfort, it is of importance to the farmer, as it allows grass to grow where it otherwise could not.—

(DIARY)

In the 1839 *Journal*, Darwin expands on his brief observation that most Australian trees do not shed their leaves annually. In comparing Australia with England, he becomes quite poetic, and shows himself to be a true Anglophile:

'This line of road is of course the best & longest time inhabited in the Colony.'

This sketch of Penrith Road was made by Martens three weeks before Darwin travelled along the same road.

The leaves are not shed periodically: this character appears common to the entire southern hemisphere, namely, South America, Australia, and the Cape of Good Hope. The inhabitants of this hemisphere and of the intertropical regions, thus lose perhaps one of the most glorious, though to our eyes common, spectacles in the world,— the first bursting into full foliage of the leafless tree. They may, however, say that we pay dearly for our spectacle, by having the land covered with mere naked skeletons for so many months. This is too true; but our senses thus acquire a keen relish for the exquisite green of the spring, which the eyes of those living within the tropics, sated during the long year with the gorgeous productions of those glowing climates, can never experience.

The Diary continues:

The greater number of the trees, with the exception of some of the Blue Gum's, do not attain a large size; they grow tolerably straight & stand well apart. It is singular, that the bark of some of them annually falls or hangs in long shreds, which swing about with the wind; & hence the trees look desolate & untidy.— Nowhere is an appearance of verdure & fertility, but rather that of arid sterility:— I cannot imagine a more complete contrast in every respect than the forest of Valdivia or Chiloe, with the woods of Australia.

 Although this is such a flourishing country, the appearance of infertility is to a certain extent the truth; the soil without doubt is good, but there is so great a deficiency in rain & running water, that it cannot produce much.— The Agricultural crops & indeed often those in Gardens, are estimated to fail once in three years; & it has so happened on more than one successive year:— So that the Colony cannot supply itself with the bread & vegetables which its inhabitants consume.— It is essentially pastoral, & chiefly so for sheep & not the larger animals: the Alluvial land near Emu ferry is some of the best cultivated which I have seen; & certainly the scenery on the banks of the Nepean, bounded to the West by the Blue Mountains, was pleasing even to the eye of a person thinking of England.

Apart from these scenic appraisals, Darwin showed his considerable and continuing interest in geology by making detailed observations in his field notebook.[17] Among the entries for the first day were the following notes:

Slight irregularities in the Stratification: Sandstone generally moderately hard; thinly stratified, on coast dip inwards — near the Nepean first meet pebbles of [coarsely] Crystalline Trappean rock & [Siliceous] Sandstones

Appears to descend to valley by steps; plain at base composed of fine Alluvial [strongly] sandy soil stratified, lying on a coarse conglomerate of above pebbles: great Escarpement of Blue Mountains.

These notes clearly illustrate that Darwin was very much the scientific observer. In combination with records made on subsequent days, they formed the basis for his formal description of the geology of New South Wales, which was published in 1844, as chapter VII of *Volcanic Islands*.

It was towards the end of the first day of his journey that Darwin encountered a group of Australian Aborigines. For a young man of twenty-six, he already had considerable experience with the native inhabitants of several countries. He had, for

'The bark of some of them annually falls or hangs in long shreds, which swing about with the wind.'

example, been an eye-witness to FitzRoy's ill-fated attempt to introduce 'civilization' to the Fuegians. He had also seen how the natives of Tahiti and New Zealand were coping with the influx of Europeans. As the following account from the Diary shows, he was quite impressed by the Australian Aborigines:

> At Sunset, by my good fortune a party of a score of the Aboriginal Blacks passed by, each carrying in their accustomed manner a bundle of Spears & other weapons.— By giving a leading young man a shilling they were easily detained & threw their spears for my amusement.— They were all partly clothed & several could speak a little English: their countenances were good-humoured & pleasant & they appeared far from the degraded beings as usually represented.— In their own arts they are admirable: a cap being fixed at 30 yards distance, they transfixed it with the spear, delivered by the throwing stick, with the rapidity of an arrow from the bow of a practised Archer: In tracking animals & men they show most wonderful sagacity & I heard many of their remarks, which shewed considerable acuteness.— They will not however cultivate the ground, or even take the trouble of keeping flocks of sheep, which have been offered them; or build houses & remain stationary.— Never the less, they appear to me to stand, some few degrees higher in civilization, or more correctly a few lower in barbarism, than the Fuegians.—
>
> It is very curious thus to see in the midst of a civilized people, savages, although harmless, wandering about without knowing where they will sleep & gaining their livelihood by hunting in the woods—
>
> Their numbers have rapidly decreased, during my whole ride, with the exception of some boys, brought up in the houses, I saw only one other party.— They were rather more numerous & not so well clothed.—
>
> I should have mentioned that in addition to their state of independence of the Whites, the different tribes go to war. In an engagement which took place lately, the parties, very singularly, chose the centre of the village of Bathurst as the place of engagement: the conquered party took refuge in the Barracks.—
>
> The decrease in numbers must be owing to the drinking of Spirits, the European diseases, even the milder ones of which such as the Measles are very destructive, & the gradual extinction of the wild animals. It is said, that from the wandering life of these people, constantly great numbers of their children die in very early life: When the difficulty in procuring food is checked of course the population must be repressed in a manner almost instantaneous compared to what can take place in civilized life, where the father may add to his labor without destroying his offspring.

In the 1839 *Journal*, Darwin clarifies this last sentence, replacing it with the following words:

As the difficulty of procuring food increases, so must their wandering habits; and hence the population, without any apparent deaths from famine, is repressed in a manner extremely sudden compared to what happens in civilized countries, where the father may add to his labour, without destroying his offspring.

Although this passage is actually concerned with the effects of a wandering life-style on child survival, Darwin is here touching on an idea that has something in common with the ideas of Thomas Malthus. In his *Essay on the Principle of Population*,[18] Malthus discussed the implications of population growth outstripping the means of subsistence. Darwin later claimed that it was from Malthus that he developed the idea of the struggle for existence, which was a key element in his theory of evolution. The appearance of the sentence on child mortality in the Diary, which was written in 1836, is significant because, although Malthus's ideas were widely known even before Darwin was born, available evidence suggests that Darwin did not read about them until September 1838.[19] From the passage in the Diary, it appears that Darwin was already thinking about the general notion of the struggle for existence at least two and a half years before reading Malthus.

In the 1839 *Journal*, Darwin also adds the following footnote to his brief comment about the measles:

It is remarkable how the same disease is modified in different climates. At the little island of St. Helena, the introduction of scarlet fever is dreaded as a plague. In some countries, foreigners and natives are as differently affected by certain contagious disorders, as if they had been different animals; of which fact some instances have occurred in Chile; and, according to Humboldt, in Mexico. (Polit. Essay on Kingdom of New Spain, vol. iv.)

In the 1839 *Journal*, he then embarks upon a lengthy discussion of the tragic effects of European settlement on native populations:

Besides these several evident causes of destruction, there appears to be some more mysterious agency generally at work. Wherever the European has trod, death seems to pursue the aboriginal. We may look to the wide extent of the Americas, Polynesia, the Cape of Good Hope, and Australia, and we shall find the same result. Nor is it the white man alone, that thus acts the destroyer; the Polynesian of Malay extraction has in parts of the East Indian archipelago, thus driven before him the dark-coloured native. The varieties of man seem to act on each other; in the same way as different species of animals—the stronger always extirpating the weaker.

Like all the entries in the 1839 *Journal*, this passage was probably written no later than June 1837,[20] and was actually printed early in 1838;[21] publication was delayed until the other two volumes, prepared by FitzRoy, were ready. The above passage indicates, therefore, that by mid-1837, Darwin was already seriously considering the implication of competition between populations.

His 1839 *Journal* continues:

> *It was melancholy at New Zealand to hear the fine energetic natives saying, they knew the land was doomed to pass from their children. Every one has heard of the inexplicable reduction of the population in the beautiful and healthy island of Tahiti since the date of Captain Cook's voyages: although in that case we might have expected it would have been otherwise; for infanticide, which formerly prevailed to so extraordinary a degree, has ceased, and the murderous wars have become less frequent.*
>
> *The Rev. J. Williams, in his interesting work (Narrative of Missionary Enterprise, p. 282.), says, that the first intercourse between natives and Europeans, "is invariably attended with the introduction of fever, dysentery, or some other disease, which carries off numbers of the people." Again he affirms, "It is certainly a fact, which cannot be controverted, that most of the diseases which have raged in the islands during my residence there, have been introduced by ships; and what renders this fact remarkable is, that there might be no appearance of disease among the crew of the ship, which conveyed this destructive importation." This statement is not quite so extraordinary as it at first appears; for several cases are on record of the most malignant fevers having broken out, although the parties themselves, who were the cause, were not affected. In the early part of the reign of George III., a prisoner who had been confined in a dungeon, was taken in a coach with four constables before a magistrate; and, although the man himself was not ill, the four constables died from a short putrid fever; but the contagion extended to no others. From these facts it would almost appear as if, the effluvium of one set of men shut up for some time together, was poisonous when inhaled by others (and perhaps more so, if the men be of different races). Mysterious as this circumstance appears to be, it is not more surprising than that the body of one's fellow-creature, directly after death, and before putrefaction has commenced, should often be of so deleterious a quality, that the mere puncture from an instrument used in its dissection should prove fatal.*

In a footnote to the discussion of the introduction of diseases by ships, Darwin cites some examples:

Captain Beechey (chap. iv., vol. i.) states that the inhabitants of Pitcairn Island are firmly convinced that after the arrival of every ship they suffer cutaneous and other disorders. Captain Beechey attributes this to the change of diet during the time of the visit. Dr. Macculloch (Western Isles, vol. ii., p. 32) says, "It is asserted, that on the arrival of a stranger (at St. Kilda) all the inhabitants, in the common phraseology, catch a cold." Dr. Macculloch considers the whole case, although often previously affirmed, as ludicrous. He adds, however, that "the question was put by us to the inhabitants who unanimously agreed in the story." In Vancouver's Voyage, there is a somewhat similar statement with respect to Otaheite: nor are these (as I believe) the only instances. Humboldt (Polit. Essay on King. of New Spain, vol. iv.) says, that the great epidemics at Panama and Callao are "marked" by the arrival of ships from Chile, because the people from that temperate region, first experience the fatal effects of the torrid zones. I may add, that I have heard it stated in Shropshire, that sheep, which have been imported from vessels, although themselves in a healthy condition, if placed in the same fold with others, frequently produce sickness in the flock.

It must be remembered that this discussion was written at a time when no one knew of the role of micro-organisms in disease. Another twenty-eight years were to pass before Pasteur published his discoveries, and it was to be even longer before the causes of disease were fully appreciated.

At 6 o'clock the next morning, Darwin crossed the Nepean River on the Emu Ferry, just a hundred or so metres upstream from the present road bridge. In his Diary, he records:

17th Early in the morning we crossed the Nepean in a ferry boat. This river, although at this spot it is both broard & deep, has a very small body of moving water.

Judging from the Martens sketch of this ferry, it was not a particularly sophisticated craft. According to the 1835 *Calendar*, the dues payable were 2d. 'for every foot passenger', and 6d. for 'every horse, mare, gelding, ass or mule, drawing or not drawing', with 'double tolls demandable on Sundays'. Since January 17 was a Sunday, and since Darwin's party consisted of two men and two horses, the total cost to Darwin was 2s. 8d.

His Diary continues:

Having crossed a low piece of land on the other side, we reached the slope of the Blue Mountains. The ascent is not steep, the road having been cut, with much care, along the side of some Sandstone cliffs:

'*Early in the morning we crossed the Nepean in a ferry boat.*'

Martens drew this sketch of the Emu Ferry eight months before Darwin used it.

A modern view of the scene depicted in Martens's sketch of the Emu Ferry.

The road described by Darwin was Major Mitchell's new road up the Blue Mountains. As a replacement for the old Bathurst road that went up Lapstone Hill, Mitchell's new road had been opened by Governor Bourke on 22 March 1834.[22] It wound its way up Mitchell's Pass, crossing the graceful stone arch bridge completed by David Lennox on 28 June 1833.[23] Both the pass and the bridge continued in use as part of the main road over the Blue Mountains for nearly one hundred years, until the present road further to the south was opened in 1926. The bridge, which is now the oldest on the Australian mainland, has been restored and is once again open to traffic, if in only one direction – eastwards or down the mountain.[24]

A map drawn by Surveyor-General Major Mitchell, showing his new line of road (to the left), which replaced the old road up Lapstone Hill. The new road later became known as Mitchell's Pass.

The Diary continues:

at no great elevation we come to a tolerably level plain, which almost imperceptibly rises to the Westward, till at last its height exceeds three thousand ft [900 metres].

By the term Blue Mountains, & hearing of their absolute elevation, I had expected a bold chain crossing the country, instead of this a sloping plain presents merely an inconsiderable front to the low country.— From this first slope, the view of the extensive woodland towards the coast, was interesting & the trees grew bold & lofty; but when once on the Sandstone platform, the scenery became exceedingly

'The ascent is not steep, the road having been cut, with much care, along the side of some Sandstone cliffs.'

Mitchell's Pass was less than three years old when Darwin travelled along it. Martens sketched the scene on 25 April 1835, and this painting was finished in the same year.

GREAT WESTERN ROAD.

34½ On crossing the ferry at Emu, the road continues across the alluvial flat of Emu Plains. A township has been planned on the rising ground overlooking the plain, and beneath this reserve, along the river bank each way, the land has been divided into farms of fifty acres each.

35 On the left, road leading to Edinglassie, and to the Government farming establishment.

36½ The road ascends by the new Pass lately made, to avoid that of Lapstone Hill, and reaches at

38½ Pilgrim Inn, one of the cleanest and most convenient in the country: it is at the junction of the old Blue Mountain road with the present ascent by the new Pass.

43½ Inn lately built by Mr. Alexander Fraser: the spot is called Fitzgerald's valley. This valley extends each northerly, and being good feed for cattle, is likewise reserved as a resting-place. There is also an old military station here, called Springwood.

Beyond Springwood the good soil disappears, and the range is poor and rugged.

Nothing breaks the monotony of this mountain road for many miles, save one or two small huts along the road, at watering places. The road itself is, in those parts where it has been formed, tolerably good; but unmade parts intervene, which are difficult for carriages. The ironed-gangs which formed the new pass are now at work on a new cut, which is to obviate the very worst of these, namely, "*Seventeen mile pinch*," by a very level and much shorter road than the present.

49 A pile of stones called Caley's repulse, well known in the history of this road.

51 Twenty-mile hollow, where there is a good spring of water, and the land is reserved as a resting place. There is also a hut, kept by Pembroke, on the right.

54 Twenty-four-mile hollow, another hut on the left.

57 On the left, King's table land.

58½ Weatherboard Hut, a very good mountain inn, on a fine stream which forms a cataract at a short distance to the southward. The water which rises in this valley, named Jamison's Valley by Governor Macquarie, is inconsiderable, but the wild scenery of the inaccessible valley into which it vanishes, is well worth the traveller's attention. The climate at this elevated station is usually very different from that of Sydney. Here the Sydney traveller no longer complains of *dyspepsia*;

" Oh there is sweetness in the mountain air !
" And health, which bloated ease was never known to share."

61 On the left, a point called Govett's Point, projecting into the valley.

64 Pulpit Hill; the land is reserved for a resting-place for cattle. From Springwood to this place, the country on either side consists of impassable gullies, descending on the right, to the river Grose, and on the left to Cox's River, above its confluence with the Wolondilly.

66 A hut.

69 Blackheath; a new inn, kept by Gardner. A mile or two to the north-east is another fine cataract, discovered by Assistant Surveyor Govett.

The direction of the road changes towards the N.W.

The 1835 *Calendar* itinerary and a modern map of the route taken by Darwin on 17 January from Emu Ferry to Blackheath. Darwin's description of the scenery along the mountain road as being 'exceedingly monotonous' is in good agreement with the description given in this itinerary.

monotonous. On each side there is a scrubby wood of small trees of the never-failing Gum family: there are no houses or cultivated land with the exception of two or three small Inns.— The road is solitary, the most frequent object being a bullock waggon piled up with bales of Wool.—

In the middle of the day, we baited our horses at a little Inn called the Weatherboard. The country here is elevated 2800 ft [850 metres] above the Sea.

When William Cox was constructing the first road over the Blue Mountains in 1814, he built a small weatherboard storage hut on the eastern side of a little stream that is now called Jamison Creek.[25] Although the hut was destroyed by fire a few years later, the area became known as Weatherboard Hut or simply Weatherboard.

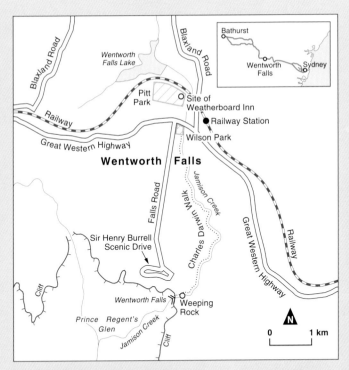

Map of present-day Wentworth Falls. The Darwin tree, marking the site of the Weatherboard Inn, is in Pitt Park. Jamison Creek rises in Wentworth Falls Lake, flows very near the site of the Weatherboard Inn, and then flows southwards across the plateau until it descends into the Jamison Valley at the Wentworth Falls. The Charles Darwin Walk starts in Wilson Park and follows the creek down to the falls.

This evergreen oak was planted by W.W. Froggatt, president of the Field Naturalists' Club, NSW, in January 1936, on the site of the Weatherboard Inn in present-day Pitt Park, Wentworth Falls. It commemorates the centenary of Darwin's visit. In the foreground is a gravel pit which exposed part of the foundations of the inn.

The latter name continued in general use for many years until gradually, from the 1870s onwards, it was replaced by Wentworth Falls. Some distance from the site of the hut, on the opposite side of the creek, an inn was built by a man named John Mills some time between June 1826 and February 1829. At the time of Darwin's visit, it was owned by William Boyles, and the licensee was Thomas Readford.[26]

The inn no longer stands, but its site is marked by the 'Darwin tree' – a Mediterranean evergreen oak (*Quercus ilex*) that was planted in January 1936, by Mr W. W. Froggatt, president of the Field Naturalists' Club, NSW, to commemorate the centenary of Darwin's visit.[27] The Darwin tree can be found in the north-eastern corner of Pitt Park, just south of the railway line.

The itinerary in the 1835 *Calendar* records that the Weatherboard Inn is located 'on a fine stream which forms a cataract at a short distance to the southward. The water which rises in this valley . . . is inconsiderable, but the wild scenery of the inaccessible valley into which it vanishes, is well worth the traveller's attention.' Having probably read this passage, and very likely having heard of the valley from his guide, Darwin left the inn and walked southwards along Jamison Creek to where it cascades into Prince Regent's Glen at what is now called the Wentworth Falls. The scene greatly impressed him:

A modern view of the 'tiny rill of water' – Jamison Creek – along which Darwin walked.

. . . *walked mile & ½ to see Cascade: most magnificent astounding & unique view, small valley not lead to expect such scene.*

. . .

Certainly most stupendous cliffs I have ever seen.

These notebook entries were expanded into the following striking account in his Diary:

About a mile & [a] half from this place there is a view, exceedingly well worth visiting. Following down a little valley & its tiny rill of water, suddenly & without any preparation, through the trees, which border the pathway, an immense gulf is seen at the depth of perhaps 1500 ft [450 metres] beneath ones feet. Walking a few yards farther, one stands on the brink of a great precipice. Below is the grand bay or gulf, for I know not what other name to give it, thickly covered with forest. The point of

*'one stands on the brink of a vast precipice. Below is the grand bay or gulf
. . . the line of cliff diverges away on each side, showing headland, behind
headland, as on a bold Sea coast.'*

Darwin wrote at great length about the 'grand amphitheatre' that encloses
the Jamison Valley, and was so impressed that he came back to view the scene
for a second time on his return journey from Bathurst. This undated Martens
painting is an excellent illustration of the scene which so excited Darwin, and
which has remained unchanged to the present day.

A modern view of the
scene described by
Darwin. In essence,
nothing has changed.

view is situated as it were at the head of the Bay; for the line of cliff diverges away on each side, showing headland, behind headland, as on a bold Sea coast.

These cliffs are composed of horizontal strata of whitish Sandstone; & so absolutely vertical are they, that in many places, a person standing on the edge & throwing a stone can see it strike the trees in the abyss below: so unbroken is the line, that it is said to be necessary to go round a distance of 16 miles [26 kilometres] in order to reach the foot of the waterfall of this little rill.— In front of the gulf & about 5 miles [8 kilometres] distant another line of cliff runs & so can have the appearance of completely encircling it; Hence the name of Bay is justified as applied to this grand amphitheatrical depression.—

If we may imagine & I believe such nearly the actual origin, a harbor & its various arms, its deep water surrounded by cliffy shores, suddenly to be laid dry; let a forest spring up on the sandy bottom, & we shall have the appearance & structure which is here exhibited. The class of view was to me quite novel & certainly magnificent.

Modern visitors to Wentworth Falls usually approach the Jamison Valley along Falls Road, and catch their first, and often only, sight of the valley and falls from one of the lookouts in Burrell Drive. While the view from these lookouts is certainly magnificent, it is not nearly as dramatic as the view that Darwin saw and described so vividly. In order to experience Darwin's view, it is necessary to approach the falls as Darwin did, by walking southwards along Jamison Creek.

For those who are fit and able, a 2-kilometre walking track called the Charles Darwin Walk was cleared alongside the creek in 1986, starting from a wooden archway in Wilson Park just south of the Great Western Highway, and finishing at the falls. Those who are less inclined to physical activity can drive to the end of Falls Road. Then, avoiding the temptation of the nearby lookouts, they should follow a path that leads down to the creek, joining the Charles Darwin Walk near a beautiful small waterfall called Weeping Rock. A hundred or so metres downstream from Weeping Rock, the path rounds what appears to be just another corner in the small meandering valley of the little creek; but instead, as Darwin describes, 'suddenly & without any preparation, through the trees, which border the pathway, an immense gulf is seen at the depth of perhaps 1500 ft beneath ones feet'.

On 18 January 1986, as part of a celebration of the 150th anniversary of Darwin's visit to Wentworth Falls, a plaque was erected on a rock beside Jamison Creek, at a spot very near where Darwin rounded the last corner and saw the magnificent view.[28]

Perhaps the most pleasing aspect of the view from this vantage point is that the scenery is still essentially the same as described by Darwin; little has changed in more than 170 years.

Aware that he still had quite a distance to travel before dark, Darwin left this 'magnificent' view, walked back to the Weatherboard Inn, and rode on to Blackheath:

> *In the evening we reached the Blackheath; the Sandstone plateau has here attained the elevation of 3411 ft [1040 metres]; & is as before, covered with one monotonous wood.— On the road, there were occasional glimpses of a profound valley, of the same character as the one described; but from the steepness & depth of its sides, the bottom was scarcely ever to be seen.—*
>
> *The Blackheath is a very comfortable inn, kept by an old Soldier; it reminded me of the Inns in North Wales. I was surprised to find that here, at the distance of more than 70 miles from Sydney, they could make up 15 beds for travellers.—*

(DIARY)

The 'very comfortable inn' in which Darwin stayed was the Scotch Thistle, which opened for business on 11 July 1831, with Andrew Gardiner (commonly misspelt Gardner) as the owner and licensee. While it probably suited Gardiner to

The Inn at Blackheath. Darwin stayed in the building with the skillion roof. The two-storey portion to the right was built in 1879, and the original building was demolished in 1938

be regarded by his guests as 'an old Soldier', the fact is that he arrived in Australia as a convict, having been sentenced at Oxford in 1818 to seven years' transportation. Like many of his fellow convicts, he prospered after serving his sentence. Within two years, he received a grant of 640 acres (259 hectares) of grazing land on the road to Bathurst, just to the west of the Blue Mountains. Two years later, he was granted 20 acres (8 hectares) on which to build an inn at Blackheath. By the time he died at the age of 74, he was the owner or lessee of large tracts of grazing land to the west of the Blue Mountains.

During a chequered history of more than one hundred years, the inn in which Darwin stayed underwent many alterations and additions, and was given many

A modern view of Gardner's Inn. The site of the building in which
Darwin stayed is at the left-hand end of the present building.

different names including Govett's Leap Hotel, Hydora Hotel, Hydora House, and Hotel Astoria. Finally, in 1938, the original building was demolished, the remaining buildings were substantially altered, and the name was changed to Gardner's Inn, in honour of the first owner and licensee. Some of the original sandstone blocks are preserved in a wall along the Bathurst Road next to the current hotel, and in the boundary wall of the nearby Baby Health Centre. Others can be found in a commemorative cairn opposite the Blackheath Railway Station.[29]

In January 1986, at the time of the sesquicentenary of Darwin's visit, a commemorative dinner was held in Gardner's Inn.[30]

Despite his long ride and walk the day before, Darwin woke early the next morning, and went on an even longer walk to see another grand view:

> *18th Very early in the morning, I walked about 3 miles to see Govett's leap; a view of a similar, but even perhaps more stupendous, character: So early in the day the gulf was filled with a thin blue haze, which, although destroying the general effect, added to the apparent depth of the forest below, from the country on which we were standing. Mr. Martens, who was formerly in the Beagle & now resides in Sydney, has made striking & beautiful pictures from these two views.—*

As we have already seen, Darwin's former shipmate Conrad Martens had arrived in Sydney in April 1835. Martens kept an account book of the pictures he painted in Australia. The first page of this book shows that by the end of August of that year, he had already completed eight paintings, including one entitled *Govets Leap* (*sic*) and another called *Fall at Weatherboard*. Darwin's comment indicates that during his visit to Sydney, he saw these two paintings.

While passing through the Blue Mountains, Darwin filled several pages of his notebook with comments on geology. He was particularly interested in the

> *Valleys, most extraordinary heads, like arm of sea, appears quite out of proportion to little streams in great fissures cleared by the sea — It is an immense Sandstone formation.*

A little further on, he writes:

> *Valley most extraordinary, expand into great depression surrounded by absolutely ⊥° cliffs. — The Chart will give correct idea of peninsula & Islands of the grand plain: in [Crest] or points of plains, notches of opposite valleys: ∴ not present causes: forms that of marine bays. Sea could not excavate Sand banks, . . . current cleavage & stratification edge of Blue Mountain? Cracks too large? Elevation acted upon by the sea?*

These hastily written notes indicate that as soon as he saw the great valleys of the Blue Mountains, he began to speculate on how they had been formed. Presumably because he had so many questions and no satisfactory answers, he omitted these sections when he wrote up his Diary. He did not, however, forget the questions. Eight years later, when his account of the geology of Australia was published as chapter VII of *Volcanic Islands*, he expanded on the entries in the notebook, and attempted to explain how the valleys were formed. The most important passages from chapter VII of *Volcanic Islands* were reproduced in the 1845 *Journal*. He starts by describing the immense valleys:

> *These valleys, which so long presented an insuperable barrier to the attempts of the most enterprising of the colonists to reach the interior, are most remarkable. Great arm-like bays, expanding at their upper ends, often branch from the main valleys and penetrate the sandstone platform; on the other hand, the platform often sends promontories into the valleys, and even leaves in them great, almost insulated, masses. To descend into some of these valleys, it is necessary to go round twenty miles [30 kilometres]; and into others, the surveyors have only lately penetrated, and the colonists have not yet been able to drive in their cattle.*
>
> *But the most remarkable feature in their structure is, that although several miles wide at their heads, they generally contract towards their mouths to such a degree as to become impassable. The Surveyor-General, Sir T. Mitchell,* endeavoured*

'Very early in the morning, I walked 3 miles to see Govett's leap; a view of a similar, but even perhaps more stupendous, character.'

in vain, first walking and then by crawling between the great fallen fragments of sandstone, to ascend through the gorge by which the river Grose joins the Nepean; yet the valley of the Grose in its upper part, as I saw, forms a magnificent level basin some miles in width, and is on all sides surrounded by cliffs, the summits of which are believed to be nowhere less than 3000 feet [900 metres] above the level of the sea.

When cattle are driven into the valley of the Wolgan by a path (which I descended), partly natural and partly made by the owner of the land, they cannot escape; for this valley is in every other part surrounded by perpendicular cliffs, and eight miles lower down, it contracts from an average width of half a mile, to a mere chasm, impassable to man or beast.

Sir T. Mitchell states that the great valley of the Cox river with all its branches, contracts, where it unites with the Nepean, into a gorge 2200 yards [2000 metres] in width, and about 1000 feet [300 metres] in depth. Other similar cases might also have been added.

(1845 JOURNAL)

The asterisk in the above passage indicates a footnote in which Darwin refers to page 154 of volume I of Mitchell's book *Travels in Australia*, and then adds the following acknowledgement:

I must express my obligation to Sir T. Mitchell, for several interesting personal communications, on the subject of these great valleys of New South Wales.

It is not known whether Darwin and Mitchell met during Darwin's visit, but as indicated in Chapter 6, they certainly met and corresponded after Darwin had returned to England. One of the letters from Darwin to Mitchell is reproduced on page 177.

Continuing with the extract from his geological account, Darwin attempts to explain how these magnificent valleys were formed. At first he suggests (correctly) that the valleys were produced by erosion:

The first impression, on seeing the correspondence of the horizontal strata on each side of these valleys and great amphitheatrical depressions, is that they have been hollowed out, like other valleys, by the action of water; but when one reflects on the enormous amount of stone, which on this view must have been removed through mere gorges or chasms, one is led to ask whether these spaces may not have subsided. But considering the form of the irregularly branching valleys, and of the narrow promontories projecting into them from the platforms, we are compelled to abandon this notion. To attribute these hollows to the present alluvial action would be preposterous; nor does the drainage from the summit-level always fall, as I remarked near the Weatherboard, into the head of these valleys, but into one side of their bay-like recesses.

(1845 JOURNAL)

Having convinced himself that his initial suggestion is 'preposterous', he then considers another possible explanation:

Some of the inhabitants remarked to me that they never viewed one of these bay-like recesses, with the headlands receding on both hands, without being struck with their resemblance to a bold sea-coast. This is certainly the case; moreover, on the

'*The valley of the Grose . . . is on all sides surrounded by cliffs.*'

present coast of New South Wales, the numerous, fine, widely-branching harbours, which are generally connected with the sea by a narrow mouth worn through the sandstone coast-cliffs, varying from one mile in width [1600 metres] to a quarter of a mile [400 metres], present a likeness, though on a miniature scale, to the great valleys of the interior. But then immediately occurs the startling difficulty, why has the sea worn out these great, though circumscribed depressions on a wide platform, and left mere gorges at the openings, through which the whole vast amount of triturated matter must have been carried away?

The only light I can throw upon this enigma, is by remarking that banks of the most irregular forms appear to be now forming in some areas, as in parts of the West Indies and in the Red Sea, and that their sides are exceedingly steep. Such banks, I have been led to suppose, have been formed by sediment heaped by strong currents on an irregular bottom. That in some cases the sea, instead of spreading out sediment in a uniform sheet, heaps it round submarine rocks and islands, it is hardly possible to doubt, after examining the charts of the West Indies; and that the waves have power to form high and precipitous cliffs, even in land-locked harbours, I have noticed in many parts of South America.

To apply these ideas to the sandstone platforms of New South Wales, I imagine that the strata were heaped by the action of strong currents, and of the undulations of an open sea, on an irregular bottom; and that the valley-like spaces thus left unfilled had their steeply sloping flanks worn into cliffs, during a slow elevation of the land; the worn-down sandstone being removed, either at the time when the narrow gorges were cut by the retreating sea, or subsequently by alluvial action.

(1845 JOURNAL)

In finally favouring, though somewhat reluctantly, this marine denudation explanation, Darwin was simply following his mentor Charles Lyell who in 1837 had introduced such an explanation for the origin of valleys into the fifth edition of his extremely influential book *Principles of Geology.*

Darwin could not accept the erosion explanation because he could not believe that so much sandstone could have been eroded away through the narrow necks of the valleys. From other sections of the 1845 *Journal*, and from other works of his that had been published by this time, it is evident that he already had quite a good understanding of the length of time over which geological events took place. However, it appears he still did not fully appreciate the geological time span involved in the formation of the valleys of the Blue Mountains; if he had realised that the erosion of the sandstone platforms has been continuing ever since they first began to rise about 180 million years ago, he might have been more inclined to accept his initial (correct) explanation.

During the decades following the writing of the above passages, Darwin, like most geologists, gradually changed his mind and eventually came to favour erosion as the primary cause of valleys such as those in the Blue Mountains.[31]

We now return to Darwin's Diary, and continue with his journey:

> *A short time after leaving the Blackheath, we descended (about 800 ft) from the Sandstone platform, by the pass of Mount Victoria. To effect this pass, an enormous quantity of stone has been cut through, the design, & its manner of execution, would have been worthy of a line of road in England, even that of Holyhead.—*

Within two years of taking up his position as Surveyor-General in 1828, Major T. L. Mitchell had marked out a new line of road extending from the western edge of the Blue Mountains to Bathurst. The first and most difficult part of this new road involved the construction of a route down the steep slopes of Mount Victoria. Overcoming the opposition of the Governor and the Colonial Secretary, both of whom preferred a different route on which construction had already commenced, Mitchell pushed ahead with his Pass of Victoria, which he considered to be 'the crowning glory of his road'. Opened on 23 October 1832, it is still in use today as one of the two main routes over the Blue Mountains.[32]

Darwin's Diary continues with an account of his journey along Mitchell's road through the beautiful granite country of the Vale of Clwydd:

> *We now entered upon a Granite country: with the change of rock, the vegetation improved; the trees were both finer & stood further apart, & the pasture between them was slightly greener & rather more abundant.—*
>
> *At Hassan's Walls, I left the high road & made a short detour from the road, to a place called Walerawang; to the superintendent of this I had a letter of introduction from the owner in Sydney.*

The 'high road' was Surveyor-General Mitchell's road, which had been in use since 1830.[33] It is still in use today as the road from Old Bowenfels to Mt Lambie via Rydal. In taking his detour to Wallerawang, Darwin followed a route very similar to that taken today by the Great Western Highway between Old Bowenfels and Wallerawang.

Wallerawang was one of the first large-scale pastoral properties to be established west of the Blue Mountains. The original property consisted of 2000 acres (810 hectares) occupied by, and then granted to, a free settler called James Walker, soon after he arrived from Scotland in September 1823. By the time of Darwin's visit twelve years later, the home property was well established, and Walker had obtained other grants of land further west. Eight years after Darwin's visit, Walker had control of twenty-seven stations throughout New South Wales, covering a total of 155 000 hectares, for which he paid a single licence fee of only £10![34]

The property later became known as Barton Park, named after Edwin Barton, the engineer-in-chief on the Zig Zag railway at Lithgow, who married Walker's daughter Georgina.[35] In 1979, the whole site of the Barton Park homestead disappeared beneath the waters of Wallace Lake. This was formed by the building of a dam across Cox's River, as a source of water for the nearby power station. In 2006, the Governor of New South Wales, Professor Marie Bashir, unveiled a plaque on a Darwin memorial beside the lake.[36]

'A short time after leaving the Blackheath, we descended (about 800 ft) from the Sandstone platform, by the pass of Mount Victoria.'

71 Leave the old road on the right, and descend by Mount Victoria, which was opened by Governor Bourke on his way to Bathurst, in October 1832. This cuts off a considerable detour by the old road, which was also excessively steep descending to the vale of Clywd, at an inclination of fifteen degrees, or about one in four; and up this acclivity the Bathurst oxen drew heavy teams, even when it was deep in mud. Verdant vallies now extend before the traveller, and present a striking contrast to the scenery of the mountains.

The surrounding country is named the vale of Clywd.

The fate of Collit's Inn, now left several miles out of the road, proves in a striking manner, the necessity for determining the true lines for Great roads in the first instance; and it may be presumed on the other hand, that the formation of such establishments along the new line to Bathurst, and the locations thereon, will, with the certainty of its being a permanent road, be proportionally more rapid.

It had long been a desideratum how to get off these mountains into the smiling valley below;—Art has at length triumphed over nature, and it is chiefly in passes such as these, that the traveller learns to appreciate justly, the "Transportation System" of Britain; by contrasting what is thus accomplished by the mere labour of her criminals, for the colonization of this fifth Continent, with the deeds of the most victorious armies of other nations—the destruction of cities, and the devastation of empires.

72¼ Bridge or platform across the head of Butler's rivulet, the rocky precipices overhanging this valley on the North, terminate abruptly in the point called Mount York, at which point, the old or first made road descended

" In many an airy whirl."

75 Cross the old road by Mount Blaxland. On descending Mount Blaxland, the old road passes the creeks, known as Jock's Bridge and Antonio's Creek; and after another steep range reaches the Fish river bridge, a difficult ford; from this part to Bathurst, the banks of the Fish River are located; the road continues by Emu Valley, Lowe's Swamp, Sidmouth Valley, and Rainville; recross the Fish River at O'Conell Plain, and onwards to Bathurst bridge, on the river Lett; a never-failing little stream

gurgling over a pebbly bed, and under shady brushes, which keep it cool and preserve it from evaporation in summer: here a village reserve has been marked.

78 Cross Major Lockyer's road and pass under the rocks which, from their resemblance to stupendous ruins in the midst of perfect solitude, may have suggested, when this road was first marked out, the name of Hassan's Walls, in allusion either to Hassan's Walls in the *Arabian Nights*, or to those lines of Byron, in his description of the "Ruin'd Palace of Hassan" in the *Giaour*—

" On desert sands 't were joy to scan
The rudest step of fellow man,
So here the very voice of grief
Might wake an echo like relief."

The noble poet has here well expressed the ideas of one lost in the bush; as Major Mitchell was near this place during a whole day when he had been but a few weeks in the Colony. The new road is carried round under these pinnacles so as to avoid the deep hollow which they enclose, and which is so deep that the curve formed by the new road is not much greater *horizontally* than that which the before-mentioned road forms *vertically* in passing right through it.

79¼ Bridge called Monaghan's Downfall, in allusion to the desperate death of a prisoner, who fell when the road was forming there.

80 Recross the road of Major Lockyer, and that to Mudgee and Talbragar, by Wallerawang, turns to the right.

The 1835 *Calendar* itinerary and a modern map of the route followed by Darwin on 18 January 1836, between Victoria Pass and the Wallerawang turnoff.

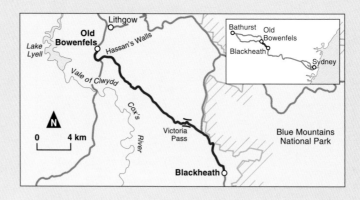

The superintendent of Wallerawang, to whom Darwin had a letter of introduction, was Andrew Brown[37] (often spelt with an 'e'):

> *I found Mr Browne a sensible well informed Scotchman; he asked me to stay the en-*
> *suing day, which I had much pleasure in doing.*
>
> *This place is a specimen of one of the large farming establishments of the*
> *Colony; it would however be more appropriately called a sheep-grazing establish-*
> *ment. They have here rather more Cattle & horses than what is common on account*
> *of some of the valleys being swampy & producing some right sort of pasture. The*
> *number of the sheep is 15,000; far the greater part of them are feeding at the distance*
> *of more than a hundred miles, under the care of different shepherds & beyond the*

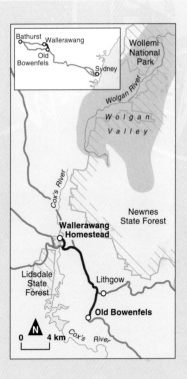

ROAD TO MUDGEE AND PANDORA'S PASS.

80¼ On the left leave the Great Western Road.
82¼ Farmer's Creek, a branch of Cox's River, immediately
    under Mount Walker.
86 Cross another of the sources of Cox's River.
88 Wallerawang, the establishment of James Walker, Esq.,
    J.P., where a road to Bathurst, as proposed by Sir Thomas
    Brisbane, turns off to the left. To the north-west of
    Wallerawang is a swamp, which forms the head of Cox's
    River; and immediately across the dividing range, to the
    eastward, is one of the heads of the second branch of the
    Hawkesbury;—this stream takes its course through a re-
    markable valley named the Wolgan, where Mr. Walker
    has his cattle. There is a very narrow pass, shut up by a
    gate, within which the cattle are secure, although within a
    range of many miles; the lower part of the valley being
    walled in by perpendicular rocks, and the outlet shut up
    by the debris of the surrounding mountains.

*'At Hassan's Walls, I left the high road & made a short detour from
the road, to a place called Walerawang.'*

The description of the Wolgan Valley given in the 1835 *Calendar*
itinerary is very similar to that given by Darwin. The modern map
shows the route of Darwin's detour to Wallerawang, and the location
of the Wolgan Valley.

limits of this Colony on unoccupied ground. They had just finished this day, the last of the shearing of 7000 sheep; the rest are sheared in another place.— I believe the value of a quantity of wool from 15,000 sheep would be more than 5000£ sterling.

Two or three flat pieces of ground, near the house were cleared & cultivated with Corn, which the Harvest men were now reaping. No more wheat is sown, than sufficient for the annual support of the labourers; the general number of assigned Convict servants is here about 40; but at present there were rather more.

Although the farm is well stocked with every requisite, there was an apparent absence of comfort; and not even one woman resided here.— The Sunset of a fine day will generally cast an air of happy contentment on any scene; but here the brightest tints on the woods surrounding this retired farm-house, could not make me forget that forty hardened profligate men were ceasing from their daily labours, like the Slaves from Africa, yet without their just claim for compassion.

(DIARY)

Darwin spent the evenings of 18 and 19 January 1836 at Wallerawang, which was then a sheep station. The homestead now lies beneath the waters of Wallace Lake, which was formed in 1979 to supply water to a local power station.

Darwin had seen many slaves during his travels in South America, and had expressed much sympathy with them. Convicts, however, were another matter. Despite an earlier Diary comment that the parties of convicts working on the roads had committed only 'some trifling offence', and by implication therefore deserved some sympathy, Darwin considers that the convicts on the farm have no 'just claim for compassion'.

Perhaps the Wallerawang convicts were a particularly tough bunch. The records of proceedings at the local courthouse at Hartley certainly lend some credence to this possibility.[38] They show, for example, that in the seven months following Darwin's visit, the superintendent Andrew Brown and his assistant David Archer were in court at least three times, testifying against the behaviour of convicts on the station: one convict was convicted of attacking an employee with an axe and a shovel after being caught killing a sheep; another was convicted of threatening the life of an overseer, and stealing his bed and some clothing, having several years earlier harboured bushrangers on the property; and a third was convicted of attempting to assault another overseer, in addition to 'losing' 23 sheep.

Darwin had probably seen some of these convicts at first hand. In addition, he recorded in his notebook that Andrew Brown had given him a

> bad account of men, not reformation, or punishment, not happy; but do not quarrel, excepting when drunk, quite impossible to reform.

With all this fresh in his mind, Darwin was led to conclude that the Wallerawang convicts were 'hardened profligate' men, who had achieved their present unhappy state solely because of their own antisocial acts. It remains an open question whether their actions were in any way a response to unjust treatment on the station.

The Diary continues:

> 19th Early on the next morning Mr. Archer, (the joint superintendent, & the only other free man about the farm) took me out Kangaroo hunting. We continued riding the greater part of the day; but my usual ill-fortune in sporting followed us & we did not see a Kangaroo or even a wild dog.—

At this stage of his life, the young and very active Darwin was still a keen devotee of the 'sport' of hunting: he had participated in many shooting parties in England, and had hunted rheas in South America. As the *Beagle*'s voyage continued, however, he gradually 'discovered, though unconsciously & insensibly, that the pleasure of observing & reasoning was a much higher one than that of skill & sport. The primeval instincts of the barbarian slowly yielded to the acquired tastes of the civilized man.'[39]

Despite the lack of kangaroos, the excursion presented Darwin with an equally interesting example of Australia's native fauna:

The Grey-hounds pursued a Kangaroo Rat into a hollow tree from out of which we dragged it: it is an animal, as big as a Rabbit, but with the figure of a Kangaroo.

(DIARY)

This was his first opportunity to examine closely an Australian marsupial. The rat-kangaroo or potoroo (*Potorous tridactylus*) is one of the oldest known of Australian quadrupeds, having been first described in 1792. Its range extends from

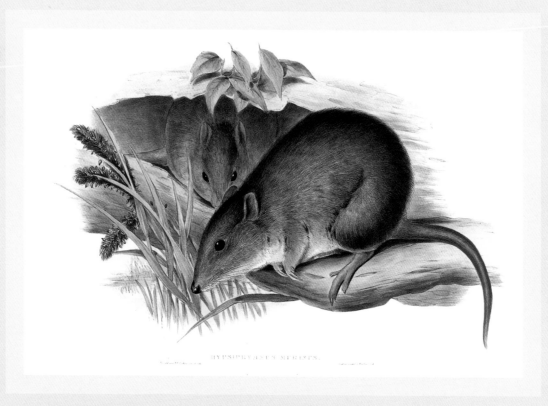

A rat-kangaroo or potoroo, as illustrated in John Gould's *The Mammals of Australia* (1845–63). The modern scientific name is *Potorous tridactylus*. Darwin caught one near Wallerawang on 19 January 1836.

south-eastern Queensland to south-western Victoria and Tasmania.[40] Since it was already so well known, Darwin felt no need to record any details about it, nor to add this animal to his collection of specimens to be sent to England.

He did, however, record his impressions about the effect on native animals of European settlement in general, and of the English greyhound in particular:

> *A few years since this country abounded with wild animals; [now] the Emu is banished to a long distance & the Kangaroo is become scarce. To both of them the English Greyhound is most destructive.*

(DIARY)

In the 1839 *Journal*, he added:

> *It may be long before these animals are altogether exterminated, but their doom is fixed.*

His dire prediction of extinction has already come true for twenty-seven species of Australian mammals, including the broad-faced potoroo (a close relative of the rat-kangaroo seen by Darwin), several species of wombats, bandicoots, wallabies, and mice, and also the Tasmanian tiger.[41] And at least another thirty-seven species of Australia mammals are endangered, along with a further seventy-two species of other Australian animals.[42] Fortunately, many other species still survive in adequate numbers.

Besides expressing his concern for the future of the native animals, Darwin was also worried about the future of the Australian Aborigines. 'Blinded' by the short-term 'trifling advantages' offered by European settlers, the Aborigines seemed certain to be driven out of most of the land to which they belonged:

> *The Native Blacks constantly are trying to borrow their dogs; their use of them, offal when an animal is killed, & milk from the Cows, are the peace offerings of the Settlers, who push, further & further inland.— The thoughtless Aboriginal, blinded by these trifling advantages, is delighted at the approach of the White man, who is doomed to rob him of his country:—*

(DIARY)

Darwin now continues recounting the day's excursion:

> *Although with bad sport, we enjoyed a pleasant ride: The woodland is generally so open, that a person on horseback can gallop through it; it is traversed by a few flat bottomed valleys, which are green & free from trees; in such spots, the scenery was like that of a Park & pretty:—*

(DIARY)

Although he does not mention it in his Diary, it is evident from his notebook that during the day's ride, Darwin and his companions went as far north as the Wolgan Valley. His notebook records that he collected at least seven different samples of rocks during the ride, including coal from a layer 'nearly a foot thick' (30 centimetres), and some slate containing fossil leaves.

Just as with his earlier reactions to the valleys at the Weatherboard and at Blackheath, Darwin was struck by the magnitude and shape of the Wolgan Valley, and was puzzled as to how it could have been formed. Page after page of the notebook entries for Tuesday 19 January consists of descriptions and speculations about this particular valley:

> *At Wolgan: grand valley surrounded by Cliffs of Sandstone . . . large ordinary form, like a bay with arms: so precipitous tho with [labour] in one spot a cattle track has been [cut] down, generally vertical walls, many hundred feet high: sides perpendicular reaching to general level of country, about 7 miles long, & one mile broard — appears to have all been removed ⌐ form modelled by water: yet exit of the valley, is by a narrow creek, a few hundred yards wide, with stupendous vertical sides, no cattle can pass [out] & twice the Surveyors have attempted to pass down the head of the small river, but have failed.*

Although none of this found its way into the Diary, it did, as with the notes taken earlier, form the basis of his long geological discourse in the 1845 *Journal*. In agreement with these notes which indicate that Darwin did visit the Wolgan Valley, readers may have noted that in the middle of the 1845 geological discourse, Darwin states that he descended the cattle path into the Wolgan Valley (see page 57). From his notebook, we now know that this occurred during the kangaroo hunt at Wallerawang.

In describing the countryside near Wallerawang, he touches upon another familiar aspect of the Australian landscape – the effect of bushfires. Wherever there were trees, there was also evidence of fire, and the only relief from the monotony of blackened stumps was guessing whether the fire had been recent or not:

> *In the whole country I scarcely saw a place, without the marks of fire; whether these may be more or less recent, whether the stumps are more or less black, is the greatest change, which breaks the universal monotony that wearies the eyes of a traveller.—*

(DIARY)

The monotony was occasionally broken by the sighting of birds:

In his notebook, Darwin describes seeing 'magnificent parrots'. He was most likely referring to the Crimson Rosella, *Platycercus elegans* (above left); the Eastern Rosella, *Platycercus eximius* (above right); or the Australian King Parrot, *Alisterus scapularis* (left). From John Gould's *The Birds of Australia* (1848).

Above left: Sulphur-crested Cockatoo, *Cacatua galerita*. Above right:
Australian Raven, *Corvus coronoides*. Below left: Australian magpie,
Gymnorhina tibicen. Below right: Pied Currawong, *Strepera graculina*.

Neither in these woods are there many birds; although certainly some of the Parrots are excessively beautiful.— I saw some large flocks of the white Cockatoo which were feeding in a Corn field; & plenty of Crows, like our jack daws, & another bird, something like the magpie.

(DIARY)

During his first day in the Blue Mountains, Darwin recorded in his notebook that he had seen some 'pretty birds' and 'magnificent parrots' just before reaching the Weatherboard Inn. He did not, however, mention these in his Diary at that time. His notebook record of birds sighted at Wallerawang consists only of 'White Coccatoos & Crows', and so provides no information beyond that given in the Diary.

Darwin's first encounter with parrots in their natural habitat was in South America, where he had seen at least two different species.[43] Now, in Australia, he renewed his acquaintance with the group. The 'excessively beautiful' ones he saw during his trip to Bathurst were most likely the Crimson Rosella, *Platycercus elegans*, the Eastern Rosella, *Platycercus eximius*, and the Australian King Parrot, *Alisterus scapularis* – all very colourful examples of Australian parrots that can still be seen in the Blue Mountains and Wallerawang districts in the summer.[44]

The white cockatoos were undoubtedly Sulphur-crested Cockatoos, *Cacatua galerita*. As Darwin observed, 'large flocks' of these birds are often seen in corn (i.e. wheat) fields. They are so fond of wheat grains (both freshly sown ones and those ripening in heads) that many Australian wheat farmers consider them pests.[45] The 'crows' that he saw were most likely Australian Ravens, *Corvus coronoides*. These birds have many features in common with the large number of other species of typical crows that occur around the world, including the Jackdaw, *Corvus monedula*, with which Darwin was familiar.[46]

The 'bird, something like a magpie' could have been either the Australian Magpie, *Gymnorhina tibicen* (often called the Black-backed Magpie) or the Pied Currawong, *Strepera graculina*. Both of these species are 'something like' the northern hemisphere Magpie, *Pica pica*.[47]

While thinking about the names that had been given to various Australian species, Darwin comments:

The English have not been very particular in giving names to the productions of Australia; one family (Casuarina) of trees are called Oaks, for no one reason, without it is, that there is no one point of resemblance;[48] animals are called tigers & Hyaenas, simply because they are Carnivorous & so on.—

(DIARY)

The day's ride having ended, the party returned to the Wallerawang homestead, situated just a short walk from Cox's River, whose shady banks must have
offered welcome relief from the midsummer heat. The next entry in Darwin's
notebook states:

> *In the evening walked up; Cox's river: chain of pools. so dry a country:— Saw several
> Ornithorhynchus: like water rats, in movements & habits.*

This was Darwin's first and only encounter with the platypus, an animal even
more unusual than the rat-kangaroo he had seen earlier in the day. His Diary gives
a fuller account:

*'In the dusk of the evening, I took a stroll along a chain of ponds
(which in this dry country represents the course of a river).'*

In his notebook, Darwin recorded that the chain of ponds was in fact
Cox's River, illustrated here in a watercolour painted in the late 1830s
by Conrad Martens.

In the dusk of the evening, I took a stroll along a chain of ponds (which in this dry country represents the course of a river) & had the good fortune to see several of the famous Platypus or Ornithorhyncus paradoxicus.[49] They were diving & playing in the water; but very little of their bodies were visible, so that they only appeared like so many water Rats. M[r]. *Browne shot one; certainly it is a most extraordinary animal; the mounted Specimens do not convey a proper idea of the head & beak; the latter being contracted & hardened.—*

This strange animal has been the subject of much interest and not a little controversy, since it was first discovered on the Hawkesbury River near Sydney in

Darwin saw several platypuses (*Ornithorhynchus anatinus*) in the Cox's River, on the evening of 19 January 1836. From John Gould's *The Mammals of Australia* (1845–63).

1797.[50] With a snout shaped like a duck's bill, a flattened body covered with seal-like fur, a broad flat tail, and webbed feet, its external appearance was sufficiently unusual to cause observers in England to believe that they were being hoaxed when the first specimens were brought from Australia. Like many other people, Darwin was surprised to discover that although the snout looks like a duck's bill, and becomes hard in mounted specimens, it is actually quite soft and pliable in the living animal. This is because it is composed of naked, thick skin spread over a framework of bone.

Darwin was not the only one to liken the platypus to the European water rat, *Arvicola terrestris* (more commonly called water vole), the species immortalised by Kenneth Grahame in *The Wind in the Willows*. Writing in 1845, the naturalist John Gould stated that 'In many of its habits and actions, and in much of its economy, the *Ornithorhynchus* assimilates very closely to the Common Water Vole . . . ; frequenting as it does similar situations, climbing stumps of trees and snags which lie prostrate in the beds of rivers, and burrowing in the bank side in an upward direction, a retreat to which it resorts during the day or on the approach of danger'.[51] Despite the similarities in adaptation, the two animals are extremely different in appearance and in physiology; the platypus is a monotreme (subclass Prototheria; order Mono-tremata; family Ornithorhynchidae) whereas the water vole is a (placental) rodent (subclass Eutheria; order Rodentia; family Muridae).

At the time of Darwin's visit, there was still a mystery concerning how the platypus reproduced. Claims by certain Aborigines that the platypus laid eggs had been dismissed as being far too fanciful, and it was not until 1884 that an English zoologist named William Caldwell confirmed the Aborigines' story during a field trip on the Burnett River in Queensland.

It seems that Darwin did not attempt to examine any burrows, and so gave himself no chance of discovering the remarkable truth about platypus reproduction. Perhaps he had heard that even though the nests are built less than 60 centimetres below the surface of the ground, they are located at the end of winding burrows that are usually between 4 and 18 metres long;[52] and Darwin's companions at Wallerawang would have lacked the Aborigines' knowledge of how to locate nests without digging out the whole burrow.[53]

However, even if he had been able to examine a nest, it is not very likely that he would have found any eggs. Mating generally occurs in the spring, most often in September or October; eggs are laid 12 to 15 days later, and hatch after a further 10 to 12 days.[54] In the normal course of events, therefore, there would be little chance of discovering any eggs in mid-January.

The next entry in Darwin's Diary is probably the best known and most discussed of all the entries that he recorded about Australia. In 1836 he wrote:

> *Earlier in the evening I had been lying on a sunny bank & was reflecting on the strange character of the Animals of this country as compared to the rest of the World. A Disbeliever in everything beyond his own reason, might exclaim, "Surely two distinct Creators must have been [at] work; their object however has been the same & certainly in each case the end is complete".—*
>
> *Whilst thus thinking, I observed the conical pitfall of a Lion-Ant:— a fly fell in & immediately disappeared; then came a large but unwary Ant; His strugglles to escape being very violent, the little jets of sand, described by Kirby (Vol I P 425)[55] were promptly directed against him.— His fate however was better than that of the poor fly's:— Without doubt this predacious Larva belongs to the same genus, but to a different species from the European one — Now what would the Disbeliever say to this? Would any two workmen ever hit on so beautiful, so simple & yet so artificial a*

'*I observed the conical pitfall of a Lion-Ant.*'

An Australian ant-lion, which is actually the larva of a lacewing (family Myrmeleontidae). The adult female lays her eggs in dry soil, and, after hatching, the larva burrows backwards into the soil, forming a conical pit into which its prey falls.

contrivance? I cannot think so.— The one hand has worked over the whole world.—
A Geologist perhaps would suggest, that the periods of Creation have been distinct &
remote, the one from the other; That the Creator rested in his labor.

In the margin beside the sentence beginning 'Without doubt', he wrote:

NB. The pitfall was not above half the size of the one described by Kirby.

When interpreting this passage, we must remember that in the last few hours, Darwin had seen Australian birds that resembled English birds of the same name, but which obviously belonged to different species; he had seen a miniature kangaroo the size of a European rabbit, behaving somewhat like a rabbit, darting about in the undergrowth; and, most striking of all, he had seen several platypuses, which in movement and behaviour might easily have been mistaken for European water rats, even though they were built to a completely different plan.

So it was that in the space of just a few hours in the middle of January 1836, on an isolated property in inland Australia, Darwin had been confronted with three clear illustrations of the fact that similar environments in completely different parts of the world seemed to be inhabited by animals having similar adaptations, but obviously belonging to different species. In the two most striking cases, the similarly adapted animals belonged to different genera, families, orders and subclasses as well.

To modern biologists, this phenomenon is called convergent evolution, and is seen as providing evidence of the power of natural selection as an adaptive force. To Darwin, it was a puzzle.

Of course, Darwin was not the first European to notice the resemblance in adaptation between Australian and European animals; some of the common names given to Australian animals that Darwin had complained about were simply the result of early explorers and settlers recognising the similarities. But Darwin was sufficiently inquisitive to really think about the reason for them; why would a creator bother to create two animals such as the rat-kangaroo and the rabbit, or the platypus and the water rat, which are so markedly different in basic design and in the way in which they bear and raise their young, but which occupy similar environments in two different parts of the world?

Not wishing to state his ideas directly, Darwin creates a suitably anonymous 'Disbeliever in everything beyond his own reason'. Having presumably rejected all current beliefs about the fixity of species and the existence of a supreme creator, the Disbeliever's initial reaction to 'the strange character of the Animals of this country as compared to the rest of the World' is to suppose that 'two distinct Creators must have been [at] work'; one in Australia, and the other in the rest of the world.

While lying on the banks of Cox's River and contemplating these questions, Darwin notices the pitfall of an Australian ant-lion, and observes that it uses exactly the same 'artificial . . . contrivance' to catch its prey as does its European counterpart; and yet its pitfall is less than half the size, and he assumes it to be definitely a different species. 'Now what would the Disbeliever say to this?' he asks. 'Would any two workmen ever hit on so beautiful, so simple & yet so artificial a contrivance?' Without any hesitation, Darwin replies: 'I cannot think so.' In other words, the Disbeliever must now think in terms of 'one hand' rather than two.

Despite the firm conclusions that various readers have drawn from this passage, it is in fact tantalisingly ambiguous. Given that Darwin was writing the Diary at least partly for the benefit of his friends and family back in England, many of whom still probably hoped that he would enter the ministry, this ambiguity may have been intentional.

If by the 'one hand' Darwin means the universal force of natural selection, then this is one of the earliest glimpses we have, if not *the* earliest glimpse, of the theory that he was to use much later to explain the mechanism of evolution. It is quite possible, however, to interpret the 'one hand' as referring to God the creator who 'hit on' the one 'artificial . . . contrivance' and used it on a slightly different scale in the two different hemispheres of the world. In writing for a very mixed audience, Darwin chose words that would offend no one.

The final sentence is also ambiguous. In a very tentative fashion, it raises the possibility that a geologist (Darwin in another disguise?) might suggest that a single hand worked at 'distinct & remote' periods. Containing words like 'the Creator' and 'periods of Creation', the sentence has a distinctly biblical character; from a certain perspective, it could readily be interpreted in a non-controversial manner by his family and friends back in England. It is also possible, however, to interpret this sentence as a rather subtle attempt to introduce a far-reaching and revolutionary idea: that under the single guiding hand of natural forces, different species have been 'created' at different times during the geological time span, in response to the environmental conditions prevailing at any particular time and place. Once again, it appears that Darwin has chosen his words carefully.

When preparing the Diary for publication as the 1839 *Journal*, Darwin made some editorial changes to the above passage, improving the English and clarifying a few points.[56] Some people may see hidden meaning in certain changes: for example, the first occurrence of the word 'Disbeliever' was changed to 'unbeliever', and the second to 'sceptic'; the emphatic words 'I cannot think so' were somewhat diluted to the less direct 'It cannot be thought so'; and 'whole world' was broadened to 'universe'. But none of these alterations seems to signal an important change in emphasis. The greatest alteration was the omission of the last sentence – the one about the Geologist perhaps suggesting that the periods of creation had been distinct and remote. Despite its tentative and ambiguous nature, it is possible that a family member or friend objected to this sentence, because they saw it as departing too far from a literal interpretation of the biblical account of creation. However, it is also possible that its omission was nothing more than an editorial decision.

By 1845, when the time came for a second edition of the *Journal* to be published, Darwin's ideas on evolution were much more fully developed. Indeed, most of the important elements of his evolutionary theory had already been committed to paper (but not to print) in his 230-page 'sketch' of 1844, which was an extension of a shorter abstract written in 1842.[57] Despite this, Darwin was still very reticent. In the Australian section of the 1845 *Journal*, all of the speculative sentences previously in the ant-lion passage have been omitted, and the remaining descriptive sentences have been relegated to a footnote.[58]

At first thought, it is difficult to know what to make of this. Was it because he was now so certain about his ideas, that speculative sentences were no longer necessary? Had he progressed beyond the stage of asking questions? If so, then why didn't he replace the questions with the answers? Why just simply remove the questions?

Perhaps he felt that this was neither the time nor the place for providing the answers. He was, after all, still meticulously compiling example after example to support the answers that were by now well formed in his own mind; and many more examples were to be collected during the next thirteen years before he was finally coaxed into putting his principal ideas into print. Even then, he still felt that he was not ready. Moreover, the *Journal* was primarily a travelogue, and a very popular one at that: the first (1839) edition, which was originally the third of three volumes describing the *Beagle*'s voyage, was reprinted as a separate volume in the same year and again the following year, and it was translated into German in 1844.[59]

When a different publisher, John Murray, asked him to prepare a second edition, Darwin was intent on condensing and downplaying the scientific parts, and improving the book's appeal still further to the general reader. His claim at the time that the second edition had been 'considerably improved & popularized'[60] is borne out by the sales: 5000 copies were printed in the first four years, and it remained a popular book well into the present century, with a further 31 700 copies appearing under the John Murray imprint alone, and unknown numbers being printed by other publishers, including translations into twenty-three other languages.[61]

It seems safe to conclude, therefore, that the speculative sentences in the ant-lion passage were victims of the condensation and downplaying of scientific aspects of the voyage that occurred during preparation of the 1845 *Journal*. It was no longer satisfactory to simply ask questions: he had certainly advanced beyond that stage. And yet to have provided appropriate answers would have required an expansion rather than a condensation of the scientific content. It would be best simply to remove the questions and save the answers until he had collected more examples, and was ready to publish his fully fleshed ideas in a proper scientific manner.

Another factor that probably encouraged Darwin's reticence was the remarkably hostile reaction that was then being given to the book *Vestiges of the Natural History of Creation*, published anonymously in 1844. The book was actually written by Robert Chambers, a Scottish publisher and amateur naturalist. Although Chambers managed to assemble a few sensible arguments concerning the fact (but not the mechanism) of evolution, he surrounded these with so many fanciful and/or erroneous arguments that the book was ridiculed. Moreover, some people even believed that Darwin was the author.[62] In such a highly charged atmosphere, it is hardly surprising that Darwin chose not to publicise his own evolutionary ideas.

Of course Darwin did expand some scientific passages in the 1845 *Journal*, as we have already seen with his discussion of the geology of the Blue Mountains. But as the following explanation shows, this type of expansion is what we might expect.

When the *Journal* was first published in 1839, Darwin considered himself as much a geologist as a naturalist, and the word Geology was given precedence over Natural History in the title. By 1845, his priorities had changed, and in the title of the second edition, Natural History has precedence over Geology. Although he was still very interested in geology, and had written three geological books about the voyage, geology was now more a means to an end, rather than an end in itself. The 'end' was by now very definitely the solution of the problem of evolution, and his future publications would reflect this: they would deal not with geology, but with all the information he was now gathering on natural history.

It is not surprising, therefore, that he decided to include what he had learned and concluded about the formation of the Blue Mountains: he would not be returning to that subject in any later publications.

Leaving scientific matters for the time being, Darwin now resumes his Diary with an account of the remainder of his trip to Bathurst:

> 20th *A long days ride to Bathurst; before joining the high road we followed a mere path through the forest.*

The Great Western Highway, which is the present main road to Bathurst, crosses over Cox's River very near to where Darwin saw the platypuses and the ant-lion pit. It then follows more-or-less the line of Darwin's 'mere path' to Mt Lambie, where it rejoins the line of Mitchell's 'high road'. From there to Bathurst, the Great Western Highway follows quite closely the route of Mitchell's road.

Soon after rejoining the 'high road', Darwin noted that he had crossed the highest part of the Great Dividing Range. In his geological diary,[63] he wrote:

About 9 miles from Walerawang we came to the line of watershed dividing the sources of those rivers whose courses are known from those which flow into the vast interior.

During the day's ride, he noted that:

With the exception of a few Squatters huts the country was very Solitary.

(DIARY)

188   ITINERARY.

80¾ Bridge in Bowen's Hollow; from whence the road ascends round the extremity of a long hill, continuing along the south-west side of it until it heads a little gully, and re-crosses (for the last time.)

81½ Lockyer's road, which, having taken a considerable turn northward, is made to cross Mount Kirkly, a hill on the right, and proceed southward. The new road continues onward along a granite ridge, which admits of as gradual a descent as could be found to the bank of the river Cox, which flows in the only ravine of magnitude between this and Bathurst. In order to obtain an easy descent, every advantage is taken of the alternate sides of three hills on this extremity, and the intervening necks, and we at length reach

83¾ Bridge on Farmer's Creek, near its junction with the river Cox; the course of this stream is from the north-east, and it rises in the range at about 12 miles from where it joins Cox's River: it is named in honour of a useful old horse named Farmer, which fell and broke its neck at this place when the party first crossed, marking the line of road. The road continues along the margin of the Cox, and skirting the base of Mount Walker, the highest summit of which is a cone, on which the trees have been cleared for the general survey. It is said that the natives have a superstitious aversion to this hill, from some very remarkable echoes which are to be heard about it, and which answer their *cooeys* after a considerable lapse of time.

The base of this mountain consists of trap-rock or basalt, which is apparent where the road has been cut along the river bank.

84½ On Cox's River, near where it makes its most westerly bend round the foot of Mount Walker, the hill terminates in a long extremity, low enough at the point to afford a site for a small village. This spot is now occupied by the Stockade, &c. for the iron-gangs formerly employed there. We ascend the opposite bank by a piece of road where the labours of the iron-gang were chiefly required; but after leaving the immediate bank of the river, the road is found to continue along a ridge nearly level, till it reaches the dividing range, if range it can be called at this part, being very low; this is nevertheless the ground dividing the waters of the interior from those which fall towards the sea, Cox's River being but a branch of the Nepean or Hawkesbury, to which it finds its way through the Blue Mountains, which are scooped by its tributaries into ravines of a grand and peculiar character. Eight miles higher up the Cox is Walerawang, the estate of James Walker, Esq. On the interior side we soon reach without scarcely any descent.

ITINERARY.   189

86½ Solitary Creek, a purling stream in a grassy vale. Here a village reserve has been marked, and several farms have been recently selected; amongst others that of Mrs. J. W. Fulton. The general course of Solitary Creek is south-west, until it falls into the Fish River, at 12 miles below. The road beyond approaches by a very gradual ascent, the side of Honeysuckle Hill (the summit of which has been cleared for the general survey) being abreast of the highest summit at 89 miles. Immediately beyond, the ground assumes a more fertile appearance, being clothed with better grass, and a more promising kind of timber. On Honeysuckle Hill the Surveyor-General passed one night, for the purpose of intersecting the Lighthouse at Sydney, and succeeded, the light appearing like a remote star, glimmering at regular intervals through the trees, on the intervening but distant horizon at the head of the Grose River.

This hill, although it seems of no great height above its immediate base, is, nevertheless, more elevated above the sea than any part of the Blue Mountains between it and Sydney, being situated on the dividing ground between the eastern and western waters.

91 At two miles beyond this hill the road reaches some open flats of good land, on which a clean and comfortable little Inn, kept by   , and called the Prince George, has been recently built. This appears to be the most convenient half-way station between Mount Victoria and Bathurst.

93¾ Meadow Flat, on a stream of water called Haye's Creek, and where a village reserve has been marked. In this neighbourhood there are one or two small farms.

94 Haye's station.

96½ Dixon's Creek, a small water-course falling towards the Fish River.

100 Cross a range of good land called the Badger Brush, which, from the great size of the trees cut there, and the number of sawyers at work, may be called the Pennant Hills of Bathurst. The road descends to a small rivulet under the hill, called Stony Range, where it crosses the road used originally in travelling by Walerawang to Bathurst. It is on emerging from the woods on the side of this ridge, that the traveller is first gratified with a sight of the plains of Bathurst.

100½ Bridge on stream under Stony Range. On crossing the neck of ground connecting this hill with the lower eminences extending towards the plains, the road reaches the head of Winburndale stream, and continues two miles along its left bank, passing the sheep stations of Mr. M'Kenzie and others.

105 Cross an inferior ridge called Brown's Hill, the last before the road reaches the plains of Bathurst.

Darwin was particularly taken with the terms used to describe various classes of citizens such as squatters, and explained them in his Diary:

A "squatter" is a freed or "ticket of leave" man, who builds a hut with bark in unoccupied ground, buys or steals a few animals, sells Spirits without a licence, buys stolen goods & so at last becomes rich & turns farmer: He is the horror of all his honest neighbours.— A "Crawler" is an assigned Convict, who runs away & lives how he can by labor or petty theft.— The "Bush Ranger" is an open villain, who lives by highway robbery & plunder; generally they are desperate & will sooner be killed than taken alive.— In the country it is necessary, to understand these three names, for they are in perpetual use.—

---

190   ITINERARY.

107½ The road reaches the open country.
113½ Cross the river Macquarie.
   From this neighbourhood there is a path leading northward to Mudgee. It first crosses the Winburndale rivulet, on whose banks are the grants of Mr. Samuel Terry, Wilshire, Bayley, and others ; then Clear Creek, by Suttor's grant, and so on to a place called Wiagdon, Lewis's grant ; thence across the Turon River, and many other streams, this path or road being of the worst possible description ; it joins the road to Mudgee and Tabrabooker Swamp at 131 miles. *(See Mudgee Road.)* To the east of this path is a place called the Lime Kilns ; to the west is a far better path, almost without a hill, leading by Millah Murrah, and nearly direct towards Mudgee.

On crossing the ford, a road extends to the south, through the small farms recently laid out, from the great reserve on the west bank of the Macquarie—several of these are already in cultivation. Opposite, on the east bank, near the confluence of Campbell's River (which is about ten miles from Bathurst), veterans have been located. The path from this crosses a fine and very extensive grant to the Church, reaching various stations and farms higher up the Campbell River.
114 Cross Queen Charlotte's Ponds, and enter the settlement of Bathurst. A Commandant, Police Magistrate, Clergyman, Commissariat Officer, &c., are stationed here : there is a Government House, Military Barracks, Commissariat Stores, &c. &c., the remainder of the ground having been recently laid out for a town on a regular plan.

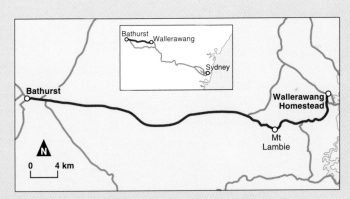

The 1835 *Calendar* itinerary and a modern map of Darwin's 'long days ride to Bathurst.'

The Diary continues with a description of the type of summer's day that is all too familiar to those who have travelled or lived in inland New South Wales at this time of the year. Darwin's description assumes added significance when it is realised that in travelling from Wallerawang to Bathurst in one day, he covered approximately 42 kilometres on horseback.

This day we had a specimen of the Sirocco-like wind of Australia; it comes from the parched interior of the Continent. Whilst riding, as always happens, I was not fully aware how exceedingly high the temperature was.— Clouds of dust were travelling in every part, & the wind felt like that which has passed over a fire.— I afterwards heard the thermometer stood at 119° [48°C] & in a room in a closed house 96° [36°C].—

In the afternoon we came in view of the downs of Bathurst. These undulating but nearly level plains are very conspicuous in this country, by being absolutely destitute of a single tree: they are covered by a very thin, brown pasture. We rode some miles across this kind of country, & then reached the township of Bathurst, seated in [the] middle of what may be called a very broard valley, or narrow plain. I had a letter of introduction to Capt Chetwode who commanded the troops there, & with him I staid the ensuing day.

Darwin's visit to Bathurst is commemorated in this plaque which is in the wall of the fernery in Machattie Park, immediately behind the courthouse. It was unveiled by Mr W.R. Glasson on 13 November 1949.

In 1836, Bathurst was a small settlement, founded by Governor Macquarie in 1815 as the first town west of the Blue Mountains. The military barracks where Darwin presumably stayed were located on the northern corner of William and Durham streets, a site currently occupied by the Bathurst Bowling Club. Under Captain Richard Chetwode's command were Lieutenant R. H. Moneypenny, three sergeants and 24 rank-and-file, all from His Majesty's 4th Foot, headquartered at Parramatta. In addition, there were eleven mounted police, consisting of one sergent, nine troopers and two dismounted troopers, under the command of Lieutenant Henry Zouch.[64]

Despite his exhausting ride, Darwin sat down that night and recorded in his notebook seven pages of comments about the geology of the region between Wallerawang and Bathurst, including notes on at least four samples of rocks that he had collected: limestone, primitive greenstone, glossy clay slate, and hornblendic greenstone.

A Martens sketch entitled *Bathurst from the West*, drawn four years
after Darwin's visit.

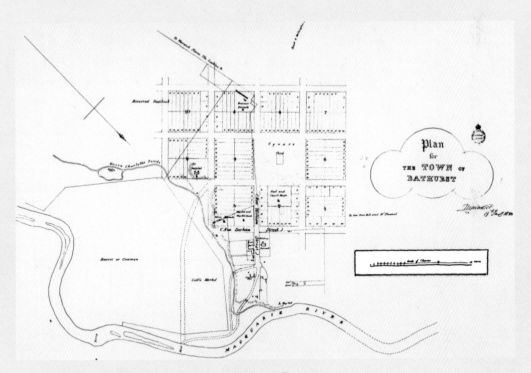

A map of Bathurst in 1833. During his visit to Bathurst, Darwin was
the guest of Captain R. Chetwode, the senior military officer in the
town. It is most likely, therefore, that Darwin stayed in the military
barracks, seen here as the northern block at the intersection of William
and Durham streets.

The next morning he wrote a letter to Captain P. P. King at Dunheved near Penrith, advising when he expected to arrive at Dunheved on his return journey. The mail-coach carrying Darwin's letter departed Bathurst at 12 noon that day, and arrived in Penrith at 6 p.m. the following day.[65]

> Bathurst
> Thursday 21st Jany 1836
>
> My dear Sir
>
>   I arrived here yesterday evening, certainly alive, but half roasted with the intense heat.— If my horses do not fail, I shall reach Dunheved on Sunday evening & if you are at home, shall have much pleasure in staying with you the ensuing day:— I have seen nothing remarkable in the Geology or indeed I may add in anything else: It appears [to] me, very singular, how very uniform the character of the scenery remains, in so many miles of country. At Mr Walker's Farm I staid one day, & went out Kangaroo hunting, but had not the good fortune even to see one. In the evening however, we went with a gun in pursuit of the Platypi & actually killed one.— I consider it a great feat, to be in at the death of so wonderful an animal.— I shall take advantage of your note of introduction to Mr Hughes & sleep there tomorrow night: if I should hear of anything remarkable in rocks of the neighbouring mountains I might be delayed there one day, in which case I should not reach Dunheved till Monday evening.—
>
>   Believe me, Dear Sir
>   Very sincerely Yours.
>   Charles Darwin.[66]

This rather unexcited letter finished, his notebook records that he 'Rode about Bathurst' for the rest of the day, but 'saw nothing'. That night, he attended a 'pleasant mess party' in the barracks.

In the next passage in his Diary, he records his impressions of Bathurst and its environs. As usual, he was very observant, and provides quite a detailed description. Unfortunately, he was not particularly impressed with what he saw:

>   Bathurst has a singular, & not very inviting appearance; groups of small houses, a few large ones, are scattered pretty thickly over 2 or 3 miles of a bare country divided into numerous fields by lines of Rails. A good many gentlemen live in the neighbourhood & some have good houses.— [67] There is a hideous little red brick Church standing by itself on a hill; Barracks & Government buildings.—

The 'hideous little red brick Church' is Holy Trinity, which still stands on the hill at Kelso, overlooking Bathurst from the east. The first church to be built west of the Blue Mountains, its foundation stone was laid in February 1834, and its first service was held on Easter Day, 23 April 1835.[68] Even its most ardent admirers would agree that when Darwin saw it soon after completion, it was not particularly attractive. One visitor, for example, later described it as 'a very ugly contrivance in itself, red brick with a little square tower and an article on the top thereof exactly like a tin extinguisher'.[69] Over the years since then, buttresses have been added, the tower has been extended, and the windows have been enlarged, with the result that Holy Trinity is now much more attractive than it was originally. As one of Australia's oldest churches, it is also now an important part of the nation's heritage.

Darwin's next Diary entry indicates a somewhat dry sense of humour – the heat and dust of an Australian summer had taken their toll:

'There is a hideous little red brick Church standing by itself on a hill.'

Darwin was not particularly impressed with Holy Trinity Church at Kelso. This photograph was taken in the 1890s, when the church was still in its original condition.

A modern view of Holy Trinity Church, Kelso, showing the extended tower, the larger windows and the buttresses, all of which have enhanced its appearance, compared with the original view that Darwin saw.

> *I was told, not to form too bad an opinion of the country, from judging of it by the road way, nor too a good one from Bathurst; in this latter respect, I did not feel the least danger of being prejudiced.*

He realised, however, that he was seeing the settlement at a particularly bad time:

> *It must be confessed; that the season had been one of great drought, & that the country does not at present wear a favourable aspect; although I understand two or three months ago it was incomparably worse.*
>
> *The secret of the rapidly growing prosperity of Bathurst is that the pasture, which appears to the stranger's eye wretched, is for sheep grazing excellent. The town stands on the banks of the Macquarie: this is one of the rivers, whose waters flow into the vast unknown interior. The North & South line of watershed, which divides the inland streams from those of the coast, has an elevation of about 3000 ft (Bathurst is 2200) & runs at a distance of about 80 or 90 miles from the sea shore.— The Macquarie figures in the maps, as a respectable river, & is the largest of those belonging to this part of the inland slope.— Yet, to my surprise, I found it a mere chain of ponds separated by almost dry land—one-from-the-other; generally a little water does flow, & sometimes there are high & most impetuous floods.— Very scanty as the quantity of the water is in this district, it becomes, further in the interior, still scarcer.—*
>
> *The Officers all seemed very weary of this place & I am not surprised at it: it must be to them a place of exile. Last year there had been plenty of Quail to shoot, but this year they have not appeared; this resource exhausted, the last tie, which bound them to existence, seemed on the point of being dissolved.— Capt. Chetwode had attempted gardening; but to see the poor parched herbs was quite heart-breaking. Yesterday's hot wind had alone cut off many scores of young apples, peaches & grapes.—*

Had he arrived in a better season, or even at another time of the year, Darwin's impressions of Bathurst might not have been so negative. But the isolated, infant settlement of Bathurst in the middle of a hot, dry summer was not likely to appeal to the young Englishman, and offered no temptation for him to stay.

RETURN FROM
BATHURST AND
IMPRESSIONS OF SYDNEY

ARWIN COMMENCED HIS RETURN JOURNEY ON

THE MORNING OF FRIDAY 22 JANUARY,

LEAVING MITCHELL'S ROAD AT KELSO, AND

FOLLOWING A MORE SOUTHERLY ROUTE.

IN HIS NOTEBOOK HE STILL ASSIDUOUSLY

RECORDED HIS GEOLOGICAL OBSERVATIONS:

(22ⁿᵈ) O'Connel plains – ⌐⌐ gravel terrace. Same class of country to Cox's river — open Woodland. almost all Granite & [illeg] Quartz & Feldspar — Which passes into a reddish Porphyry & a white Euritic ore, in several parts some dark coloured Trappean rocks

In his Diary, he wrote:

22ⁿᵈ. I commenced my return, taking a new road, called Lockyer's line.— The country was rather more hilly & picturesque.— At noon [we] baited at a farm house (there being no inns); the owner had only come out two years, & appeared to be going on very well; He had two pretty daughters, who, I suspect, will not remain long on his hands.—

Major Edmund Lockyer's road had been cleared in 1829, during his brief period as Surveyor of Roads and Bridges.[1] Very soon, however, it was superseded by Mitchell's road. As Darwin says, Lockyer's road follows a 'more hilly & picturesque' route, passing through O'Connell plains, following the Fish River to Tarana, and then continuing eastwards to Sodwalls, before joining Mitchell's road at South Bowenfels, near the western foot of Hassan's Walls.[2] The modern road from Bathurst to Bowenfels via O'Connell follows Lockyer's road for much of the way.

The Diary continues:

This was a long day's ride & the house where I meant to sleep, was off the road & not easy to find — I observed on this, & indeed all other, occassions, the general civility amongst the lower orders; when one considers what they are & have been, this is rather surprising.— The farm, where I passed the night, was owned by two young Englishmen, who had only lately come out & were beginning a Settlers life; the picture of the total want of all comfort (& in this instance filth) was not very attractive.

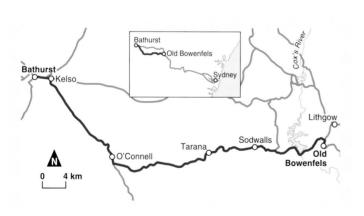

Modern map showing Darwin's return route along what was then called Lockyer's line, from Bathurst to Old Bowenfels near Hassan's Walls.

Darwin's surprise at the 'civility' with which he was treated arose from the fact that a large proportion of the people with whom he came in contact would have started life in the colony as convicts. Memories of the 'forty hardened profligate men' at Wallerawang, and of the locals' definition of a squatter, were presumably still fresh in his mind.

It is not known whether the farm where he passed the night was the one owned by Mr Hughes, to whom he had been recommended by Capt. King. Despite his mind having been broadened by some first-hand experience of country inns in South America, Darwin was appalled by the 'filth', or to use the unabridged note-book version, 'horrid filth', that he encountered while staying with the two young Englishmen.

At this point, Darwin's notebook records that he experienced yet another aspect of life during an Australian summer: 'at evening great fires'. And they were still there the next day: 'in morning raging fire'. His Diary expands slightly on these notes:

23rd *We passed through large tracts of country in flames; volumes of Smoke sweeping across the road.—*

Despite the danger so often posed to humans by bushfires, and despite being close enough to experience 'volumes of Smoke sweeping across the road', Darwin does not seem to have been in the least bit worried; he remains the dispassionate onlooker.

The Diary then gives a very brief account of his journey that day:

Early in the day we came into our former road, & ascended Mount Victoria: I slept at night at the Weatherboard, & before dark took a walk to the grand Amphitheatre.

These few words leave the reader with little idea as to what was actually involved in the day's journey. In fact, as on the previous day, Darwin and his companion had covered around 37 kilometres on horseback, firstly along the remainder of Lockyer's road until it rejoined Mitchell's road near South Bowenfels, and then back through the Vale of Clwydd, up Victoria Pass, and along the mountain road, past the Scotch Thistle at Blackheath, until they reached the Weatherboard Inn at present-day Wentworth Falls. Apparently not at all exhausted by this ride, Darwin had then walked a round trip of 5 kilometres through the bush, in order to have another look at the 'grand Amphitheatre' that had so impressed him during his outward journey.

Returning to the inn, his notebook records an intriguingly ambiguous comment about life in Australian hotels:

I do not perceive any difference in manners at the Inns from England.

Was this a criticism or a compliment? Unfortunately, we shall never know.

At this point, it is useful to summarise what Darwin had achieved to date on his inland trip. In the eight days since leaving Sydney, he and his companion had spent six days on the road, covering a total of approximately 270 kilometres[3] at an average of 45 kilometres per day. In addition, he had ridden an unknown number of kilometres during the kangaroo hunt at Wallerawang, and on the day when he 'rode about Bathurst'. On top of this, he had walked twice to see the Jamison Valley, and once to see Govett's Leap, involving a total of about 19 kilometres. And all this had been accomplished during the middle of an Australian summer, with the maximum temperature regularly near the century on the Fahrenheit scale.

It is perhaps not surprising, then, that the next entry in his notebook is simply:

24th Ill in bed.

His Diary entry for the 24th was almost equally brief:

24th In the morning, I did not feel well, & I thought it more prudent not to set out.—

As often happens in the Blue Mountains, the weather closed in overnight, and the Diary provides an evocative description of a type of day so different from those Darwin had experienced during the preceding week:

The ensuing day was one of steady drizzling rain; all was still, excepting the dropping from the eaves; the horizon of the undulating Woodland was lost in thin mist: the air was cold & comfortless — it was a day for tedious reflection.—

His notebook adds somewhat mysteriously:

Perhaps good for me | Jobs-comfort, nice girl rain for three weeks

But Darwin was impatient to move on:

26th Escaped from my prison; Having crossed the wearisome Sandstone plain, descended to Emu ferry. A few miles further on, I met Capt. King, who took me to his house at Dunheved. I spent a very pleasant afternoon walking about the farm & talking over the Natural History of T. del Fuego.

The two men had last seen each other on 27 December 1831, when King stood on the wharf at Devonport in Plymouth harbour, farewelling the *Beagle* on its present voyage. Now retired to his farm, King was busy preparing his account of the *Beagle*'s first surveying voyage. Like the present voyage, the first one had been concerned primarily with the southern coasts of South America, including, of course, that southernmost maze of inhospitable islands: Tierra del Fuego.

A view of Dunheved by Conrad Martens, painted the year after
Darwin's visit.

A sketch of Dunheved, c. 1840, probably by the owner's son and
Darwin's shipmate P. G. King (Jnr). Darwin stayed here on his return
trip from Bathurst. South Creek is in the foreground.

King had been an enthusiastic collector and observer of natural history during his *Beagle* voyage, and had already published some of his observations. Reprints of a paper describing his collection of barnacles and molluscs had reached the colony by the time of Darwin's visit, and King presented a copy to his guest, inscribing it 'Charles Darwin Esq from the Author Dunheved Jan 26 1836'.[4] It is little wonder, then, that the two men had much to discuss when 'walking about the farm & talking over the Natural History of T. del Fuego'.

P. P. King: the owner of Dunheved at the time of Darwin's visit; previously commander of the *Beagle*'s first surveying voyage; and father of the *Beagle*'s midshipman P. G. King (Jnr).

The farm itself had been established in 1806, when King's father, P. G. King (Snr), then governor of the colony, had granted his son 660 acres (267 hectares) on the banks of South Creek. Further grants by later governors extended the acreage to 4260 (1725 hectares) by 1822. Although born on Norfolk Island, King had been absent from the colony for much of this time, being educated in England, and then employed by the Royal Navy to survey the Australian coastline in four separate voyages from 1817 to 1822, and the coast of South America between 1826 and 1830. After farewelling the *Beagle* late in 1831, he had sailed for Sydney, and was by now firmly settled at Dunheved.[5]

The farm remained in the King family until 1904, and the rambling homestead remained standing until the 1950s. By this time the land had been resumed by the Commonwealth government for military purposes, and the house was demolished. The homestead site today is an area of cleared ground near the northern end of Dunheved Golf Course, on the eastern bank of South Creek, close to where the railway line crossed the creek. Fortunately, it has been designated as a preserved site, has been listed on the Register of the National Estate, and will become a public park under the control of a government authority.[6]

This was the farmhouse to which Conrad Martens had come soon after arriving in the colony in April 1835, in order to deliver his letter of introduction from Captain FitzRoy. Judging by Martens's immediate and continuing success in selling paintings to leading members of colonial society, King must have provided the help that FitzRoy had requested. Certainly King had the right connections, being the son

A watercolour by James Lethbridge Templar entitled *Beagle, an Australian Bred Horse, by Skelton, the Property of Captn P. P. King, R.N.* Apparently Captain King had such fond memories of his association with the *Beagle* that he named one of his horses after her. This painting is dated December 1839, four years after Darwin's visit.

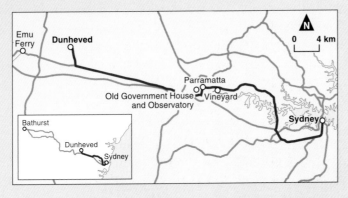

Modern map showing Darwin's return route from Emu Ferry, via Dunheved and Vineyard, to Circular Quay.

of a former governor, and related by marriage to the influential Macarthur family:
not to mention his creditable career with the Royal Navy, his Fellowship of the Royal
Society, and his active involvement with one of the colony's largest landowners, the
Australian Agricultural Company.[7]

Given these connections, it is most likely that King and his wife would have
been invited to the vice-regal ball that was held that night at Government House,
Parramatta, 24 kilometres east of Dunheved. The
reason for the ball was twofold. First, 26 January
was Anniversary Day (now called Australia Day).
On this day 48 years earlier, Captain Arthur
Phillip had raised the Union Jack in Sydney Cove,
heralding the foundation of the new colony. Earlier
in the afternoon, down in Sydney Cove, the officers
of the *Beagle* had 'dressed ship in commemoration
of the settlement being founded'.[8] In addition,
the second of David Lennox's grand bridges had
been opened that very afternoon near Liverpool,
about 26 kilometres south-west of Sydney, where
the Great South Road (now the Hume Highway)
crosses Prospect Creek. After witnessing the
Governor officially open the Lansdowne Bridge, a
'gallant show of well-dressed dames and gentlemen'
partook of 'cold collation' in a nearby marquee,
before making their way by horse and carriage to
Parramatta for the ball.[9]

Hannibal Macarthur, who
entertained Darwin to lunch
in the new house on his
property Vineyard, near
Parramatta, on 27 January
1836. Captain FitzRoy had
been entertained in a similar
manner eight days earlier.

Was Darwin invited to the ball? Perhaps he
and King decided that in the limited time they
had together, they would much prefer to relax at
Dunheved, discussing 'the Natural History of T.
del Fuego'.

The next morning, Darwin set off for Parramatta, on the second-last stage of his journey:

*27th Accompanied by Capt. King rode to Paramatta. Close to the town, his brother in law
Mr. MacArthur lives & we went there to lunch. The house would be considered a very
superior one, even in England.— There was a large party, I think about 18 in the
Dining room.— It sounded strange in my ears, to hear very nice looking young ladies
exclaim, "Oh we are Australians & know nothing about England".—*

'Close to the town [Captain King's] brother in law M^r. MacArthur lives & we went there to lunch.'

This romantic view of Hannibal and Maria Macarthur's new house on their property Vineyard, at present-day Rydalmere, was painted by Conrad Martens in 1840, four years after Captain FitzRoy and Darwin had been to lunch there. The building survived until July 1961, when, amid much public outcry, it was demolished to make way for an industrial estate.

Above: the entire painting. Below: a detail, showing the house.

The 'Mʳ. MacArthur' to whom Darwin refers was Hannibal Hawkins Macarthur, who was born at Plymouth in England in 1788. He had come to Australia in 1805, at the request of his uncle, John Macarthur, to act as a business agent for John's many enterprises in the colony. In 1812, Hannibal married P. P. King's sister Anna Maria, and the following year bought a property called Vineyard, one of the oldest farms in the colony. As we saw earlier, it was located on the northern bank of the Parramatta River, approximately one mile downstream from the town of Parramatta, just across the river from John Macarthur's Elizabeth Farm. The 'very superior' house was a grand two-storey Greek Revival mansion that was built by James Houison, probably to the design of the architect John Verge, who also designed Camden Park for John Macarthur and Elizabeth Bay House for Alexander

Front view of Vineyard, as it appeared just prior to its demolition.
The balcony was added between 1889 and 1893, and is therefore not
shown in Martens's painting.

Macleay. When Darwin went to lunch there, the house was very new: in August of the previous year, it had still been under construction.[10]

For Darwin, the contrast between the scene at Vineyard and at some of the farms he had seen during the previous week must have been very marked.

At the front of the house was a veranda with a colonnade of Doric columns, each cut from a single piece of stone; surrounding the pair of panelled front doors was a beautiful fanlight and side-lights; in the vestibule was another pair of doors, covered with baize, and also with fanlight and side panels; in the stair hall behind the inner doors stood two wooden Ionic columns, and behind them was a graceful, curved, cantilevered, stone staircase with a cedar handrail; in each of the main rooms, the fireplaces were marble.[11]

Vineyard's colonnade was saved during demolition of the house, and now stands beside the Village Green at the University of New South Wales.

Vineyard's elegant main doorway.

The house was run on a grand scale. In the dining room, lunch was served by Scottish maids, beneath a branching chandelier. The manservants were English or Indian (none were convicts, unlike many of the farm workers); a nursery-governess cared for the children; a drill-sergeant from the local regiment gave the children dancing lessons; the large cellar had quarter-casks of port, sherry and marsala on draught, and bottled beer from England; and the family maintained a holiday house at the Gap on South Head, not far from the lighthouse that Darwin had seen as the *Beagle* entered Sydney Harbour. Outside the house, bantam hens, guinea-fowl and peacocks wandered among the olive trees and roses that grew behind the new house. The children's pets, however, were distinctly Australian – native parrots, possums, and a white cockatoo.[12] The 'nice looking young ladies' included some of Hannibal and Maria's six daughters. At the time of Darwin's visit, the oldest four girls were Elizabeth, aged 20; Annie, 19; Catherine, 17; and Mary, 14. Darwin was not the only person to be attracted to these young women: his *Beagle* shipmates later married two of them. John Wickham, the *Beagle*'s first lieutenant and Darwin's close friend, married Annie in 1842; and another close friend, Philip King (Jnr), married Elizabeth in 1843.[13]

From the beginning of 1836 until November 1848, when the Macarthurs were forced to sell the property because of severe financial problems, Vineyard was one of the major centres of social life in the colony. So many officers of visiting ships were entertained by Hannibal and Maria that Vineyard became known as The Sailors' Home. They usually arrived by boat. In those days, Parramatta River was a pleasant and busy waterway, and Vineyard had its own jetty, which was only a short stroll from the house. Among the naval visitors were the officers of the *Beagle* on her next voyage (she spent a total of ten months anchored in Sydney Harbour on four separate visits between July 1838 and February 1843) and the officers of HMS *Rattlesnake*, including Captain Owen Stanley and presumably the young assistant surgeon, Thomas Henry Huxley, who was later to become Darwin's most able and vocal advocate. Other visitors included the explorers Allan Cunningham, Count Strzelecki, and Ludwig Leichhardt.[14]

When the Macarthurs were finally obliged to sell, the property was bought by the Catholic Church, which established in the house the first Benedictine Priory in Australia. Renamed Subiaco (after the Italian town near which St Benedict spent three years in a cave before establishing what became the Benedictine Order), this beautiful house was operated as a priory from 2 February 1849 until 19 December 1957, when the nuns moved to Pennant Hills to escape the noise and pollution from the factories that had gradually surrounded them.[15] In June 1961, the property was sold to one of the factories, and Subiaco was demolished during the last week of the following month, amid much public outcry. Today the site of Vineyard is a vast, sprawling industrial estate at the end of Brodie Street, Rydalmere. Although some parts of the house were saved and have been re-erected elsewhere, nothing remains of the rural property that Darwin visited, which was, in the words of the second youngest Macarthur daughter Emmeline, 'bounded on one side by a tidal-river navigable for small steamers, and on the other side by extensive forests chiefly composed of gumtrees: . . . large gardens, and, in the heart of the forest, a semicircular terraced vineyard with a stream at the foot bordered with ferns and mimosa: a lovely spot'.[16]

After what was no doubt a very satisfactory lunch in these idyllic surroundings, Darwin said goodbye to the Macarthurs and to Captain King, and departed for Sydney:

> *In the afternoon I left this most English-like house & rode by myself into Sydney.—*
>
> (DIARY)

Because Vineyard was on the Sydney side of Parramatta, his shortest route would have been to follow the road that went from Vineyard to Kissing Point (present-day Ryde), where he would have turned right into the new North Road (now Great North Road), crossing Parramatta River on a ferry at Bedlam Point (at the end of present-day Punt Road, Gladesville). After rejoining the Great Western Road (Parramatta Road) at the bridge across Iron Cove Creek, it was then simply a matter of following this road back into Sydney; and back to the *Beagle* anchored in Sydney Cove.[17]

We can assume that the officers were pleased to see him, and eager to hear of his exploits. Although they had taken shore leave, the officers' life on board the anchored *Beagle* was rather dreary, and the days passed without excitement. The main relief from being 'employed variously on Ship's duty' was taking delivery of fresh beef and vegetables.[18]

The officers were, however, able to tell Darwin one exciting piece of news: Captain FitzRoy had had an argument with his former commanding officer, Captain King. Since Darwin had spent most of the last two days with King, he probably knew of the disagreement already. But he would have been keen to hear the officers' version of events.

Although King was at Dunheved when the *Beagle* arrived, it seems that, not surprisingly, FitzRoy and King renewed their acquaintance some time during the next few days. There would have been much to discuss. King would have been particularly eager to hear how FitzRoy had fared in continuing and extending their surveys of the southern coast of South America. He would also have wished to reach an agreement with FitzRoy as to how the accounts of their two voyages should be published. During these discussions, something went wrong. In the words of P. G. King (Jnr):

During the Beagle stay in Port Jackson my Father and Capt FitzRoy had some unpleasantness about the Surveys and Charts.[19]

Had FitzRoy's notorious quick temper once again got the better of him? Whatever the reason, this argument must have been weighing heavily on FitzRoy's mind during the next few days. The evidence for this is contained in a letter written by Hannibal Macarthur to King. In order to understand the circumstances of this letter, we first have to discuss another topic.

As we saw in Chapter 1, one of FitzRoy's main tasks in Sydney was to take some of his chronometers to the observatory in the grounds of Government House.

North front and ground plan of the Parramatta Observatory, to which Captain FitzRoy took three of the *Beagle*'s chronometers, in order to check them at a site that was regarded by Captain Beaufort as being 'absolutely determined in longitude'. The observatory was located just to the west of Old Government House at Parramatta. Today, all that remains are the two mountings marked 't'.

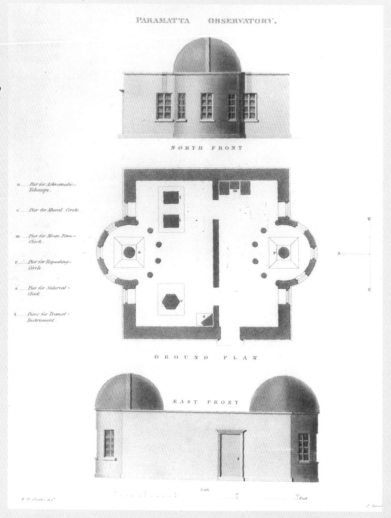

Parramatta. Since he had been instructed to establish a chain of chronometrical measurements (longitudes) around the world, it was vitally important to compare his own determinations with those few that were already firmly established. (The Sydney observatory on Observatory Hill was not built until 1858.)

It is not known exactly when FitzRoy went to Parramatta, but we do know that he took three chronometers with him,[20] that he went by what was then the most convenient route, namely by boat up the Parramatta River, and that he was accompanied by Colonel Kenneth Snodgrass, a senior member of the Legislative Council. We also know that he was in Parramatta on Monday 18 January, two days after Darwin had set out on his journey to Bathurst.[21]

That evening, there was a party at Government House, Parramatta, and FitzRoy attended.[22] So, too, did Hannibal and Maria Macarthur, and 'Grandmother' Anna King, the wife of the former Governor P. G. King and mother of Maria Macarthur and Captain King. In the following letter to King, Hannibal Macarthur describes meeting FitzRoy at the party, and relays the compliments that FitzRoy passed to him about P. G. King (Jnr), who, as we have seen, was Macarthur's nephew and King's son:

The present-day remains of the Parramatta Observatory

Vineyard
Tuesday Morning

My dear Phillip,

We spent the Evening last at Govt House as arranged and there we met Capt. Fitzroy!— Your mother had much conversation with him and he appeared much gratified but a something nervous and rather constrained was very observable — He spoke in the highest Terms of your Dear Boy and told me when I thanked him for his kindness to him — That P.G. owed to himself and his own amiable disposition all that we say about him to please us — He expressed a desire to talk on another Subject both with your Mother and Maria! regreted exceedingly that he did not come up to us with you!! said — being shut up long on board Ship rendered Men irritable and evidently he is desirous to repair the Breach. Your Sister asked him to lunch here on his way to Sydney for the purpose of his proposed conversation with your Mother and He is coming with Col. Snodgrass who came up in his Boat with him. Therefore I put off my start until the afternoon, when (D.V.) Eliz[th] *and I will come to you — I therefore despatch Essington that you may not be kept in suspense at our non-arrival.*

With kindest Complts to Mr Wickham and love to Harriet and all your Circle— I am Yours affectly

H H Macarthur

P.S. General Darling[23] *now Sir Ralph has put down his Enemies by their own Mouths! and has been invested with the G*[d]*. Cross of the Guelphic Order with strong expressions of approbation from the King —*[24]

This is a very informative letter. First, it presents the homely picture of FitzRoy and the elderly Mrs King deep in conversation about the exploits of P. G. King (Jnr); FitzRoy was full of praise for young King, and, like any proud grandmother, Mrs King was only too willing to listen to the tributes being paid to her grandson.

It also shows that FitzRoy raised the issue of his disagreement with King, albeit obliquely, by expressing 'a desire to talk on another Subject' with King's mother

Old Government House, Parramatta, as it is today, held in trust by the National Trust of Australia (NSW). Captain FitzRoy was entertained in the dining room (above right) on the evening of 18 January 1836.

and sister, and by letting it be known that he was 'desirous to repair the Breach'. It seems that Maria was only too willing to help in these delicate negotiations, for she invited FitzRoy to lunch at Vineyard the next day. Since FitzRoy would be returning to Sydney by boat along the Parramatta River, it would be a simple matter to stop off at Vineyard's jetty. So FitzRoy, accompanied by Colonel Snodgrass, had lunch at Vineyard on Tuesday 19 January.

Apparently, Macarthur had intended to travel to Dunheved to see King that morning, but he was now obliged to stay and have lunch with FitzRoy. King's son Essington, who must have been staying at Vineyard at the time, was sent instead to explain the change in plans. The fact that Macarthur sent best wishes to Wickham indicates that, like Darwin, the *Beagle's* first lieutenant visited Dunheved. In Wickham's case, he would have been renewing an acquaintance that extended back to the *Beagle's* first voyage, when Wickham had served as mate and later as lieutenant under King. The other recipient of Macarthur's best wishes was King's wife, Harriet.

The final chapter in the disagreement between FitzRoy and King is a letter to P. G. King (Jnr), written by FitzRoy from his cabin in the *Beagle*, after he had returned to Sydney:

Beagle
Monday 25th.

My dear King

I am prevented from going to Parramatta — as I intended.

Could your good Father be induced to forgive my late offences, and once more venture on board this Vessel. I give you my word that neither he, nor you, should ever regret the step.

You must send a Power of Attorney to Stilwell or some other person in England to enable him to draw your pay for you.

I hope to see you on Wednesday.

Believe me

Most faithfully your friend.

Rob[t] FitzRoy[25]

As we shall see, Captain King did accept FitzRoy's invitation, and, in the words of P. G. King (Jnr), 'the misunderstanding was happily made up'.[26]

The sentence about drawing pay in England was necessary because, as we shall also see, P. G. King (Jnr) had decided to leave the *Beagle* and stay with his family at Dunheved.

There were two days remaining before the *Beagle* sailed for Hobart, and Darwin had much to do. He had to catch up on his Diary; there were letters to write; he had to pay another visit to Conrad Martens's studio in Bridge Street to take delivery of two paintings of South America that he had decided to buy; and, with any luck, he might be able to spare a few hours to indulge in his favourite pastime – collecting beetles.

First, the Diary. It was time to take stock; time to assess his reaction to the colony:

> *28th & 29th Before we came to the Colony, the things about which I felt most interest were, the state of Society amongst the higher & Convict classes & the degree of attraction to emigrate. Of course after so very short a visit, our opinion must rank as vague conjectures; but it is as difficult not to form some opinion, as it is to form a correct judgment.—*

With this eminently sensible qualification, he started by considering the state of society in general. What would it be like living here? Would he be tempted to emigrate?

> *On the whole, from what I heard, more than from what I saw, I am disappointed in the state of Society.— The whole community is rancorously divided on every subject, into parties. Amongst those, who from their station of life ought to be amongst the best, many live in such profligacy, that respectable people cannot associate with them. There is much jealously between the rich emancipists & their children, & the free settlers.—*

In the 1839 *Journal*, he expanded on this last sentence by adding:

> *the former being pleased to consider honest men as interlopers.*

The Diary continues:

> *The whole population poor & rich are bent on acquiring wealth; the subject of wool & sheep-grazing amongst the higher orders is of preponderant interest. The very low ebb of literature is strongly marked by the emptiness of the Booksellers shops; these are inferior to the shops of the smaller country towns of England.—*
>
> *To families there are some very serious drawbacks to their comforts, the chief of which being surrounded by convict servants, must be dreadful. How disgusting to be waited on by a man, who the day before, was perhaps, by your representation flogged for some trifling misdemeanour? The female servants are of course much worse; hence children acquire, the use of such vile expressions: I heard of one instance, where the dear little innocent must have perfectly astounded its Mama.—*

On the other hand, the capital of a person will without trouble produce him treble interest as compared to England; & with [care] he is sure to grow rich. The luxuries of life are in abundance & very little dearer, as most articles of food are cheaper than in England. The climate is splendid & most healthy, but to my mind its charms are lost by the uninviting aspect of the country.

One great advantage Settlers possess is, that it is the custom to send their sons, when very young men (16–20 years) to take charge of their remote farming stations; here they directly provide for themselves; this however must happen at the expense of their boys associating entirely with convict servants.—

I am not aware that the tone of Society has yet assumed any peculiar charac-ter; but with such habits & without intellectual pursuits, it must deteriorate & become like that of the people of the United States.—

With that rather unflattering reference to the USA, Darwin reached a firm conclusion about whether he would like to settle in New South Wales:

The balance of my opinion is such, that nothing but rather severe necessity should compel me to emigrate.—

Next, he considered the colony's economic potential, again emphasising the inadequacy of his conclusions owing to his lack of understanding:

The rapid growth of prosperity in this Colony is to me, not understanding Political Economy, very puzzling.— The two main exports are Wool & Whale Oil;— Now to both of these there is a limit. The country is totally unfit for Canals; therefore there is a not very distant line, beyond which the land carriage of wool, will not render it worth while to shear & tend sheep: The pasture everywhere is so thin, that already Settlers have pushed far into the interior; moreover very far inland the country appears to become less profitable.—

I have before said, Agriculture can never succeed on a very extended scale. So that, as far as I can see, Sydney must ultimately depend, upon being the centre of commerce for the Southern Hemisphere; & perhaps on her future Manufactories:— possessing Coal, she always has the moving power at hand.—

I formerly imagined that Australia would rise into as grand & powerful a country as N. America, now it appears to me, that such future grandeur & power is very problematical.—

One hundred and seventy years on, it is perhaps surprising that any of Darwin's predictions have been realised. He was certainly off the mark in a number of areas, notably in the future of wool production. But he could see that the environment and climate of inland Australia really would impose limitations on the size of the population that could be supported, unlike the case in the USA, which has a much higher proportion of good agricultural land in reasonable environments and climates, and is therefore capable of supporting a far larger population, and thus of generating greater wealth.

In a sense, Australia is only just now having to face some of the problems foreseen by Darwin. For the last 170 years, she has been able to rely on agricultural exports to a far greater extent than Darwin envisaged. However, it is now becoming increasingly clear that she 'must ultimately depend, upon being the centre of commerce for the Southern Hemisphere; & perhaps on her future Manufactories'.

Darwin's final comments were reserved for the convicts:

> *With respect to the state of the Convicts, I had still fewer opportunities of judging, than on the other points. The first question is, whether their state is at all one of punishment; that it is not a very severe one, no one will attempt to maintain.*
>
> *The corporeal wants [of the convicts] are tolerably well supplied; their prospect of future liberty & comfort is not distant & on good conduct certain. A "ticket of leave", which makes a man as long as he keeps clear of crime & suspicion, free within a certain district, is given, upon good conduct, after years proportional to the length of the sentence: for life, eight years is the time of probation; for seven years, four, &c.—*
>
> *Yet, with all this, & overlooking the previous imprisonment & wretched passage out, I believe, the years of assignment are passed with discontent & unhappiness: & an intelligent man remarked to me, they know no pleasure beyond Sensuality.—*
>
> *The enormous bribe, which Government possesses in offering free pardons & the horror of the secluded penal Settlements, destroy confidence between the convicts & so prevents crime.— As to a sense of shame, such a feeling does not appear to be known.— It is a curious fact, but universally I was told, that the character of the convict population was that of arrant cowardice; Although not unfrequently men became desperate, & quite indifferent of their lives, yet that a plan, requiring cool or continued courage was seldom put into execution.—*
>
> *The worse feature in the whole case is, that although there is what may be called a legal reform, or that little, which the law can touch is committed, yet that any moral reform should take place, appears to be quite out of the question — I was assured by well informed people, that a man, who should try to improve, could not, while living with the other assigned servants;— his life would be one of intolerable misery & persecution.— Nor must the contamination of the Convict ships & prisons both here & in England be forgotten.—*

> *On the whole, as a place of punishment, its object is scarcely gained; as a real*
> *system of reform, this has, as perhaps would every [other plan], completely failed.—*

In the 1839 *Journal*, he added:

> *But as a means of making men outwardly honest,—of converting ragabonds most*
> *useless in one hemisphere into active citizens of another, and thus giving birth to*
> *a new and splendid country—a grand centre of civilization—it has succeeded to a*
> *degree perhaps unparalleled in history.*

In the apparently never-ending debate about the merits of free settlers as compared with ex-convicts (emancipists), Darwin leaves no doubt as to whose side he is on. It must be remembered, however, that he had seen the colony very much through the eyes of the free settlers. There is a delightful irony in the fact that when he encountered an ex-convict acting like a free settler and hiding his past, namely Andrew Gardiner at Blackheath, Darwin appears not to have suspected the man's history. He saw such men as illustrating the virtues of their present class rather than the class to which, unbeknown to Darwin, they had previously belonged.

He was prepared to admit, as we have seen earlier, that some people had become convicts for the most trifling of offences, and that perhaps they had been unfairly treated by society. But even if such people started off as 'honest', his opinion (and the opinion of many of the free settlers) was that they would soon be tainted by the others, and would soon, like the others, be beyond the reach of moral reform. Ex-convicts, therefore, were only 'outwardly honest'; they were 'vagabonds' converted into 'active citizens'.

The perplexing problem was, in Darwin's mind and in the minds of many others, that these very people were 'giving birth to a new and splendid country – a grand centre of civilization'. How could this happen?

As he pondered this question, perhaps Darwin began to realise that the world is full of 'outwardly honest' people, and that prosperous societies everywhere (including the one in England to which he was so attached), rely on such people for their continued prosperity.

Darwin wrote two letters in Sydney: one to his sister Susan, and the other to Professor Henslow in Cambridge. Inevitably, they contain descriptions of the colony and accounts of his trip to Bathurst that are similar to those already given in his Diary. So as to avoid repetition, we shall quote only those passages that contain new material.

In his letter to Susan,[27] he commenced with a brief financial report for his father:

Sydney
January 28ᵗʰ.— 1836

My dear Susan
 The day after tomorrow we shall sail from this place; but before I give any account of our proceedings, I will make an end with Business.— Will you tell my Father that I have drawn a bill for 100 £, of which Fifty went to pay this present & last year's mess money. The remaining fifty is for current expenses; or rather I grieve to say it was for such expenses: for all is nearly gone.— This is a most villainously dear place; & I stood in need of many articles.

Later in the same letter, he returns to the matter of money:

 Tell my Father I really am afraid I shall be obliged to draw a small bill at Hobart. I know my Father will say that a hint from me on such subject is worthy of as much attention, as if it was foretold by a sacred revelation. But I do not feel in truth oracular on the subject.

At least he had one good excuse:

 I have been extra\<vag\>ant & bought two water-color sketches, one of the S. Cruz river & another in T. del Fuego; 3 guineas each, from Martens, who is established as an Artist at this place. I would not have bought them if I could have guessed how expensive my ride to Bathurst turned out.

During the thirteen months that Martens had spent with the *Beagle* he had filled four books with sketches of South America. Now settled in Australia, he was busy turning some of these sketches into paintings. The two bought by Darwin show scenes that would have rekindled fond memories of times shared with Martens.

 To an unknowing outside observer, *Hauling the Boats up the Rio Santa Cruz* may not be a particularly interesting painting. But to Darwin and Martens, it recalled an exciting adventure in which twenty-five officers and crew took three whale-boats 225 kilometres up the Santa Cruz River, from its mouth on the east coast of Patagonia (southern Argentina). Battling a strong current of four to six knots, the party was divided into two shifts, and everyone, including Darwin, Martens, and FitzRoy, took turns walking along the river bank, hauling the boats upstream. At night, each person spent an hour on sentry duty, looking out for the much-feared native inhabitants. Apart from the first 50 kilometres, which had been explored by members of the previous *Beagle* expedition, this part of Patagonia was a blank on all maps.

It had taken seventeen days of hard toil to cover the 225 kilometres of the journey upstream, but only four exhilarating days for the strong current to carry them the same distance back to the mouth of the river. At three guineas, Martens's painting must have been irresistible.[28]

The second painting, called *The Beagle in Murray Narrow, Beagle Channel*, reminded Darwin of the majestic scenery of Tierra del Fuego, where the *Beagle* had spent so much time during the earlier part of the voyage. It would also have reminded him of FitzRoy's unsuccessful attempt to 'civilize' the native Fuegians, who in the painting, are approaching the *Beagle* in their small boats. Woollya, the site chosen by FitzRoy for resettlement of the three Fuegians whom he had taken, at the end of the previous *Beagle* voyage, to England for education, is very near to Murray Narrow. So too is Mount Darwin, the second highest peak in Tierra del Fuego, which, much to Darwin's delight, FitzRoy had named in his honour.

The last two entries for January 1836 in Martens's account book record the sale of two paintings to C. Darwin Esq., at a cost of three guineas each. The dates probably indicate when the paintings were completed, in good time for Darwin to take delivery when he returned from his trip to Bathurst.

Leaving money matters to one side, Darwin continues the letter to his sister with a surprisingly positive description of New South Wales:

This is really a wonderful Colony; ancient Rome, in her Imperial grandeur, would not have been ashamed of such an offspring. When my Grandfather wrote the lines of "Hope's visit to Sydney Cove" on Mr Wedgwood's medallion he prophecyed most truly.

Soon after Captain Arthur Phillip had arrived in Sydney to establish the first European settlement, he had collected some clay from Sydney Cove, and sent it to Darwin's grandfather, Josiah Wedgwood, who had recently established his pottery factory at Etruria, in the English midlands. In the words of Captain Phillip, 'Mr.

Hauling the Boats up the Rio Santa Cruz. This is one of the Martens watercolours bought by Darwin from the artist in Sydney in January 1836.

Wedgwood caused a medallion to be modelled, representing Hope, encouraging Art and Labour, under the influence of Peace, to pursue the means of giving security and happiness to the infant settlement'.[29]

Inspired by the medallion, Darwin's other grandfather, Dr Erasmus Darwin, wrote a poem – perhaps the first poem ever written about Australia. Entitled 'Visit of Hope to Sydney Cove near Botany Bay', it was published in 1789, in Phillip's account of the First Fleet's voyage to the new colony, and was reproduced by Captain FitzRoy in volume II of the *Narrative*:

> *Where Sydney Cove her lucid bosom swells,*
> *Courts her young navies and the storm repels,*

The Beagle in Murray Narrow, Beagle Channel [Tierra del Fuego]. This is the other Martens watercolour purchased by Darwin from the artist in Sydney in January 1836. In his account book, Martens called this painting *View Ponsonby Sound*.

High on a rock, amid the troubled air,
Hope stood sublime, and wav'd her golden hair;
Calm'd with her rosy smile the tossing deep,
And with sweet accents charm'd the winds to sleep;
To each wild plain, she stretch'd her snowy hand,
High-waving wood, and sea-encircled strand.
'Hear me,' she cried, 'ye rising realms! record
Time's opening scenes, and Truth's unerring word.—
There shall broad streets their stately walls extend,
The circus widen, and the crescent bend;
There ray'd from cities o'er the cultur'd land,
Shall bright canals, and solid roads expand.—
There the proud arch, Colossus-like, bestride
Yon glittering streams, and bound the chasing tide;
Embellish'd villas crown the landscape scene,

The medallion which was modelled by Darwin's grandfather,
Josiah Wedgwood, from Sydney Cove clay sent to England by
Captain Arthur Phillip. It shows Hope encouraging Art and
Labour, under the influence of Peace, to pursue the means of
giving security and happiness to the infant settlement of Sydney.

Farms wave with gold, and orchards blush between.—
There shall tall spires, and dome-capt towers ascend,
And piers and quays their massy structures blend;
While with each breeze approaching vessels glide,
And northern treasures dance on every tide!'
Here ceased the nymph—tumultuous echoes roar,
And Joy's loud voice was heard from shore to shore—
Her graceful steps descending press'd the plain;
And Peace, and Art, and Labour, join'd her train.

Darwin thought that his grandfather had 'prophecyed most truly', and FitzRoy, in the *Narrative*, proclaimed that 'all that was foretold in this allegory had come to pass, with one exception only, that of canals'.[30]

There were actually several copies of the medallion made; one of them is now in the Mitchell Library in Sydney.

In completing his letter to Susan, Darwin shows how homesick he had become by this stage of the voyage. He also gives an accurate assessment of FitzRoy's complex character:

> *From Sydney we go to Hobart Town from thence to King George Sound & then adie<u> to Australia. From Hobart town being superadded to the list of places I think we shall not reach England before September: But, thank God the Captain is as home sick as I am, & I trust he will rather grow worse than better.*
>
> *He is busy in getting his account of the voyage in a forward state for publication. From those parts, which I have seen of it, I think it will be well written, but to my taste is rather deficient in energy or vividness of description. I have been for the last 12 months on very Cordial terms with him.— He is an extra ordinary, but noble character, unfortunately however affected with strong peculiarities of temper. Of this, no man is more aware than himself, as he shows by his attempts to conquer them. I often doubt what will be his end, under many circumstances I am sure, it would be a brilliant one, under others I fear a very unhappy one.*
>
> *From K. George Sound to Isle of France, C. of Good Hope, St. Helena, Ascencion & omitting the C. Verd's on account of the unhealthy season, to the Azores & then England.— To this last stage I hourly look forward with more & more intense delight; I try to drive into my stupid head Maxims of patience & common sense, but that head is too full of affection for all of you to allow such dull personages to enter. My best love to my Father.— God Bless you all. My dearest old Granny*
>
> *Your most affectionate brother*
> *Charles Darwin.*

The other letter Darwin wrote in Sydney was to Professor Henslow,[31] with whom he had maintained a regular, mainly scientific, correspondence throughout the voyage. As with his letter to Susan, it is obvious that Darwin is by now sick of travelling – he can't wait to get back to England:

Sydney—
January— 1836

My dear Henslow,
      This is the last opportunity of communicating with you, before that joyful day when I shall reach Cambridge.— I have very little to say: But I must write if it was only to express my joy that the last year is concluded & that the present one, in which the Beagle will return, is gliding onwards.— We have all been disappointed here in not finding even a single letter; we are indeed rather before our expected time, otherwise I dare say I should have seen your handwriting.— I must feed upon the future & it is beyond bounds delightful to feel the certainty that within eight months I shall be residing once again most quietly, in Cambridge. Certainly I never was intended for a traveller; my thoughts are always rambling over past or future scenes; I cannot enjoy the present happiness, for anticipating the future; which is about as foolish as the dog who dropt the real bone for its' shadow.—

He continues with some very strong praise for the colony, seeing its rapid growth and development as a magnificent illustration of the power and skill of the British people:

You see, we are now arrived at Australia: the new Continent really is a wonderful place. Ancient Rome might have boasted of such a colony; it deserves to rank high amongst the 100 Wonders of the world, as showing the Giant force of the parent country.

And yet he would not like to live here:

On the whole I do not like new South Wales: it is without doubt an admirable place to accumulate pounds & shillings; but Heaven forfend that ever I should live, where every other man is sure to be somewhere between a petty rogue & bloodthirsty villain.—

Darwin concludes his letter with a scientific report, telling Henslow what he has done since the last letter, which was written before the *Beagle* had left the mainland of South America:

I last wrote to you from Lima, since which time I have done disgracefully little in Nat: History; or rather I should say since the Galapagos Islands, where I worked hard.— Amongst other things, I collected every plant, which I could see in flower, & as it was the flowering season I hope my collection may be of some interest to you.— I shall be very curious to know whether the Flora belongs to America, or is peculiar. I paid also much attention to the Birds, which I suspect are very curious.—

The Geology to me personally was very instructive & amusing; Craters of all sizes & forms, were studded about in every direction; some were s<uch> tiny ones, that they might be called quite Specim<en> Craters,— There were however a few facts of interest, with respect of layers of Mud or Volcanic Sandstone, which must have flowed liked streams of Lava. Likewise respecting some grand fields of Trachytic Lava.— The Trachyte contained large Crystals of glassy fractured Feldspar & the streams were naked, bare & the surface rough, as if they had flowed a week before.— I was glad to examine a kind of Lava, which I believe in recent days has not in Europe been erupted.—

These words are the first that Henslow heard of Darwin's visit to the Galapagos Islands. The plants did turn out to 'be of some interest', and the birds certainly turned out to be 'very curious'. In fact, as is now well known, the evidence that Darwin collected in the Galapagos Islands was of major importance in leading him to an understanding of how evolution had occurred.

The Galapagos report is followed by a comment about Tahiti and New Zealand, including a defence of the missionaries, who had been so roundly criticised by some people, notably Augustus Earle, the *Beagle*'s former artist:

In our passage across the Pacifick, we only touched at Tahiti & New Zealand: at neither of these places, or at sea had I much opportunity of working.— Tahiti is a most charming spot.— Every thing, which former Navigators have written is true: "A new Cytheraea has risen from the ocean". Delicious scenery, climate, manners of the people, are all in harmony. It is moreover admirable to behold what the Missionaries both here & at New Zealand have effected.— I firmly believe they are good men working for the sake of a good cause. I much suspect that those who have abused or sneered at the Missionaries, have generally been such, as were not very anxious to find the Natives moral & intelligent beings.—

After 170 years, many people would question Darwin's defence of the missionaries. Yet from his perspective, they were doing a good job. In the Tahitian section of his Diary, he had earlier listed their achievements: they had abolished human sacrifices, parricide, infanticide, warfare, idolatrous priests, and profligacy; and had greatly reduced dishonesty, licentiousness, and intemperance. Would those who criticise missionaries, he asked, prefer to be shipwrecked on an island that had been subjected to the civilising influence of missionaries, or would they prefer an island where human sacrifices were still the order of the day? In the nineteenth century, when the possibility of shipwreck was always not too far from the mind of sailors, this argument had considerable appeal.[32]

The letter finishes with a lament about what lies ahead in the voyage:

> *During the remainder of our voyage, we shall only visit places generally acknowledged as civilized & nearly all under the British Flag. There will be a poor field for Nat: History & without it, I have lately discovered that the pleasure of seeing new places is as nothing. I must return to my old resource & think of the future, but that I may not become more prosy I will say Farewell, till the day arrives, when I shall see my Master in Natural History & can tell him, how grateful I feel for his kindness & friendship.*

> *Believe me, Dear Henslow*
> *Ever yours Most Faithfully*
> *Chas. Darwin*

In fact, as we shall see, the 'field for Nat: History' was not as barren as he feared. What he really meant was that since the remainder of the trip would take him mainly to places already well established as British colonies, he would be seeing only areas that had already been well explored by Europeans, and so he would be less likely to make discoveries as exciting as those he had made in the relatively unexplored territories that the *Beagle* had visited earlier in her voyage.

These apparent limitations did not stop Darwin from investigating the natural history of Australia. While in Sydney, for example, his zoological diary reveals that he or Covington collected an

> *Oyster: small pool: Muddy almost separated from the sea. New S. Wales. Jany. 1836.*

and some

> *Shells, living, in a Muddy Salt Water pool, almost separated from the sea. Same locality as the Oyster in Spirits (1356). Sydney. Jan-y*

Also included in the list of Sydney specimens are a mud whelk, several air breathers, a sand snail, and a trochid or top shell, plus a crab, a snake and various frogs and lizards (all of which had been described previously).[33]

The sole mammal amongst the Sydney specimens, a mouse, turned out to be previously undescribed. Initially called *Mus gouldii* when named by George Waterhouse in 1837, it was later renamed *Pseudomys gouldii*. Unfortunately, it is now extinct.[34]

At some time during the *Beagle's* stay in Sydney, Darwin went beetlecollecting. He had been introduced to this pastime by his cousin, W. D. Fox, while they were both undergraduates at Cambridge, and had maintained a passion for it ever since.

Given the little time available, Darwin's expedition must have been quite short. However, Syms Covington also went insect collecting for Darwin during the *Beagle's* stay in Sydney. Darwin and Covington between them captured ninetyseven different species representing five different orders in the environs of Sydney. In a letter to Fox, written two weeks later in Hobart (see p. 147), Darwin says, 'At Sydney I took a fine species, & long did I look at it, as compared to any other insect'.[35]

Although Darwin was a keen entomologist, he was not expert at identification, and left this task to others to whom he gave his insect collection once he had returned to England. In time, many scientific papers were written by his colleagues describing the insects he had collected throughout the voyage. These papers show that of the ninetyseven species collected in Sydney, forty-two were previously unknown. Included among these new species were a leaf beetle (*Idiocephala darwini*), a seed bug (*Ontiscus darwini*), a gasteruptiid wasp (*Foenus darwinii*) and a bee (*Halictus darwiniellus*) that were each named after Darwin. The remaining novel insects comprised six Leaf beetles (Chrysomelidae), four Stink bugs (Pentatomidae), a Seed bug (Lygaeidae), an Assassin bug (Reduviidae), a Water boatman (Corixidae), a Leafhopper (Cicadellidae), a Cicada (Cicadidae),

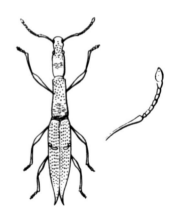

A weevil, *Leptosomus acuminatus*, one of the many beetles collected by Darwin and Covington in Sydney.

a Flatid planthopper (Flatidae), a Froghopper or Spittlebug (Cercopidae), three Parasitic wasps (Chalcididae), an Encyrtid (Encyrtidae), five Eucaratids (Eucharitidae), a Eulophid (Eulophidae), four Seed chalcids (Eurytomidae), five Lamprotatidae, and one Torymid (Torymidae).[36]

Having covered Darwin's activities, we now turn to Captain FitzRoy. During the whole of the voyage, FitzRoy maintained a regular correspondence with the Hydrographer to the Admiralty, Captain Beaufort, who had issued FitzRoy with his instructions. In a letter written in Sydney on Anniversary Day, 26 January, FitzRoy told Beaufort that 'My messmate Mr. Darwin is so much the worse for a long voyage that I am most anxious to hasten as much as possible. Others are ailing and much require that rest which can only be obtained at home'.[37]

Although it is clear from Darwin's strenuous excursion to Bathurst that he was still quite physically fit, FitzRoy's comment does help to explain Darwin's longing for England and home. In common with other members of the ship's company, Darwin was in need not so much of a physical rest as of a mental holiday from the rigours of the long voyage.

Like Darwin, FitzRoy kept a diary. Unfortunately, the Australian section is now lost. His official account of the voyage, which was published in 1839 at the same time as Darwin's *Journal*, is presumably based on his diary. FitzRoy's account of Sydney is similar to Darwin's. He was impressed with the growth and prosperity of the infant colony, but was concerned at the single-minded manner in which the citizens pursued profit to the exclusion of all else. This unhappy state of affairs was reflected in the sorry condition of the bookshops. In common with Darwin, FitzRoy was appalled by the convicts, but even more appalled by the ex-convicts.

During his stay in Sydney, FitzRoy met the Surveyor-General, Major Mitchell, who presented the *Beagle*'s captain with a copy of his recently published map (the one we assume Darwin carried on his trip to Bathurst), and a star chart, presumably of the southern hemisphere. Two days before departure, FitzRoy wrote to Mitchell:

> *Beagle*
> *Sydney Cove*
> *28 Jan.y / 36.*

My dear Sir,

I thank you very much for the Map of N.S.Wales and the Chart of the Stars which you have so kindly presented to me.

I shall take great pleasure in shewing them to my friends in England,—as proof of the very advanced state of the arts in this surprising country and as testimonials of your own unwearied skill and zeal.

Captain King has borne testimony to their extreme correctness in places with which he is acquainted;— and even I, a stranger, have heard of the many difficulties against which you have fought your way.

Will you honor me by accepting a parallel ruler made according to the suggestions of the person whose name it bears?

I hope it may find an employment in your Studio.

Prevented by most unfortunately timed indisposition — I cannot say farewell in person but most heartily do I wish you success and happiness, and a pleasanter field for your arduous exertions. Farewell my dear Sir — and believe me with much respect

Your's very Sincerely
Robt FitzRoy[38]

The exact nature of FitzRoy's 'indisposition' is not known, but it was probably the same problem that prevented him from going back to Parramatta, as mentioned in his earlier letter to P. G. King (Jnr).

As well as writing to Mitchell, FitzRoy that day took delivery of a Conrad Martens painting. Unlike Darwin, who chose two views of South America, FitzRoy selected a Tahitian scene called *View at Moorea*, for which he paid two guineas.[39] As we have seen, Martens left the *Beagle* while she was still in South America. But both Martens and the *Beagle* called at Tahiti on their separate ways across the Pacific Ocean; Martens's visit occurring several months before that of the *Beagle*.

With letter writing and all other jobs completed, it was time to leave. On the morning of Saturday 30 January 1836, the *Beagle* was released from her mooring and was then warped (hauled by rope) out of Sydney Cove. At 10.30 a.m., she began tacking her way down Port Jackson, in a strong NNE wind. Four hours later she passed through the Heads, and turned south towards Hobart. In his Diary, Darwin wrote:

> *30th. The Beagle made sail for Hobart town: Capt King & some other people accompanied us a little way out of Harbor.— Philip King remains behind & leaves the Service.—*

This passage shows that the disagreement between FitzRoy and King was finally resolved, and that King did accept FitzRoy's invitation to 'once more venture on board this vessel'. In his unpublished autobiography, King's son Philip recalls that:

> *It was determined by my Father that I should remain with him in the Colony, leaving the Service and the Shore attractions[40] made it easily for me to do so.— The Beagle accordingly sailed without me but I had an affectionate farewell from all my shipmate the Captain and M^r. Darwin.*
>
> *On settling down at Dunheved I was employed by my Father transcribing the notes of his Voyage and preparing them for publication.[41]*

The 'notes of his voyage' formed the basis of the official account of the first surveying voyage of the *Beagle*. When Philip King had done as much as he could, the manuscript was sent to England, where FitzRoy saw it through to publication, as volume I of the *Narrative*. Volume II was FitzRoy's account of the second voyage, and volume III was Darwin's *Journal*. All three volumes were published in 1839.

Three days after the Beagle's departure, Captain King wrote a brief letter to the naval hydrographer Francis Beaufort in London, reporting on the ship's visit to Sydney.[42] He took the opportunity to make some comments which help to explain FitzRoy's 'indisposition':

> *I regret to say he has suffered very much and is yet suffering much from ill health — he has had a very severe shake to his constitution which a little <u>rest</u> in England will I hope restore for he is an excellent fellow and will I am satisfied yet be a shining ornament to our service.*

In confirming that his son Philip had stayed in Sydney, King explained that:

> *he has no chance of succeeding in the service and here he will do very well*

The extent to which these predictions were fulfilled for both FitzRoy and P. G. King is discussed in Chapter 6.

The next entry in Darwin's Diary describes the *Beagle*'s approach to Hobart.

Hobart and Environs

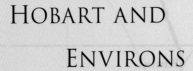

4

WHEN THE *BEAGLE* LEFT SYDNEY, THE WEATHER WAS

FINE AND SUNNY. IN THE TWO DAYS THAT

FOLLOWED, IT WAS OVERCAST AND MISTY

WITH STRONG WINDS, ALTHOUGH STILL QUITE

WARM. AS THEY NEARED TASMANIA, THE

TEMPERATURE DROPPED AND THE WEATHER

DETERIORATED:

February 5th After a six days passage, of which the first part was fine & the latter very cold & squally, we entered the mouth of Storm Bay: the weather justified this awful name.— This Bay should rather be called a deep Estuary, which receives at its head the waters of the Derwent.—

(DIARY)

In the early afternoon they passed the lighthouse called Iron Pot Light, at the southern tip of South Arm, and entered the mouth of the Derwent River. Despite the inclement weather, Darwin's servant Syms Covington made a sketch of the lighthouse from the *Beagle*'s deck.[1]

The entrance to the River Derwent, Tasmania, showing the Iron Pot lighthouse. This sketch was drawn by Darwin's servant Syms Covington from the deck of the *Beagle* as it entered the Derwent River. The lighthouse is located on the northern tip of South Arm, on the eastern side of the entrance to the Derwent River.

In his Diary, Darwin describes the scene:

Near its mouth, there are extensive Basaltic platforms, the sides of which show fine facades of columns; higher up the land becomes mountainous, & is all covered by a light wood.— The bases of these mountains, following the edges of the Bay, are cleared & cultivated; the bright yellow fields of corn, & dark green ones of Potato crops appeared very luxuriant.

As we shall see below, Darwin made many astute geological observations during his visit to Van Diemen's Land. However, like many other early naturalists and navigators, including Matthew Flinders, he was not correct when he described the cliffs at the mouth of the Derwent River as being composed of basalt: they are actually composed of a related type of rock called dolerite.[2]

At 6.40 p.m., the *Beagle* came to, furled its sails, and moored in Sullivans Cove:

Late in the evening we came to an anchor in a snug cove, on the shores of which stands the capital of Tasmania, as Van Diemen's land is now called.— The number of Ships was not very considerable.— The first aspect of the place is very inferior to that of Sydney; the latter might be called a city, this only a town.—

In the morning I walked on shore,— The streets are fine & broard; but the houses rather scattered: the shops appeared good: The town stands at the base of M. Wellington, a mountain 3100 ft, but of no picturesque beauty: from this it receives a good supply of water, a thing which is much wanted in Sydney:—

Round the Cove, there are some fine Warehouses; & on one side a small Fort — Coming from the Spanish Settlements, where such magnificent care has generally been paid to the fortifications, the means of defence in these parts appeared very contemptible.—

(DIARY)

The early settlers of Hobart had not taken their defence as seriously as the South Americans. The 'small Fort' that Darwin mentions was Mulgrave Battery, which had been established on a promontory (Battery Point) on the south side of Sullivans Cove in 1818. The fort was never used in anger, and the neighbouring land was soon put to agricultural use. Housing subdivisions commenced in 1830, and by the time Darwin arrived, the promontory looked decidedly residential.[3]

In 1831, the hill on the southern side of Sullivans Cove had been excavated sufficiently to make room for the warehouses mentioned by Darwin. These warehouses have survived, and today form a fine backdrop to the cove.[4]

Darwin continues:

Comparing this town to Sydney, I was chiefly struck with the comparative fewness of the large houses, either built or building. I should think this must indicate that fewer people are gaining large fortunes. The growth of small houses is most abundant; & the vast number of little red brick houses, scattered on the hill, behind the town, sadly destroys its picturesque effect.—

In London, I saw a Panorama of a Hobart town; the scenery was very magnificent, but unfortunately there is no resemblance to it in nature.—

(DIARY)

The waterfront warehouses in Salamanca Place are a famous landmark of Hobart. Their construction began a few years before Darwin's visit, and continued into the early 1840s.

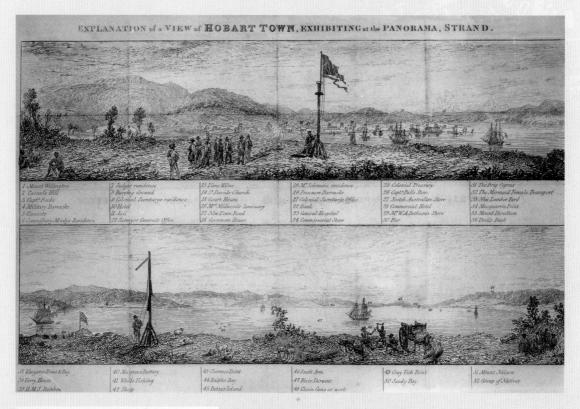

EXPLANATION of a VIEW of HOBART TOWN, EXHIBITING at the PANORAMA, STRAND.

DESCRIPTION

OF A

VIEW

OF

HOBART TOWN,

VAN DIEMAN'S LAND,

AND

THE SURROUNDING COUNTRY,

NOW

EXHIBITING

AT THE

PANORAMA, STRAND.

PAINTED BY THE PROPRIETOR,
MR. R. BURFORD.

LONDON:
PRINTED BY NICHOLS AND SONS,
EARL'S COURT, CRANBOURN STREET
1831.
PRICE SIXPENCE.

'In London, I saw a Panorama of a Hobart town.'

The panorama was prepared by R. Burford, and was exhibited by Burford in the Strand in 1831. The title page and key to the panorama are from a guidebook that was on sale at the exhibition. The viewer is standing approximately on the site where the house Secheron was later built, overlooking Secheron Point. Sandy Bay is to the right, and Battery Point (Mulgrave's Battery) is to the left. The perspective is somewhat distorted because of the inclusion of a much wider view than can be seen in real life by a person looking in any particular direction. The warehouses in Salamanca Place had not yet been built.

Panoramas were popular in England at that time. They provided a magnificent view of a far-off place that the viewers would be most unlikely to ever visit, at least voluntarily. Just before the *Beagle* sailed from England at the commencement of her voyage, Darwin had paid sixpence to see Mr R. Burford's panorama of Hobart, which was then on exhibition in the Strand.

This panorama was actually based on six watercolours painted by Darwin's former shipmate Augustus Earle. We have already seen that Earle visited Hobart briefly in 1825. Three years later, after he departed from Sydney for the last time in October 1828, his ship *Rainbow* called at Hobart on its way to India.[5] This gave Earle a second opportunity to prepare the preliminary sketches for the paintings. From Darwin's comments, it appears that he was not aware of Earle's role in the Hobart panorama. Perhaps this was because Burford did not acknowledge the artist in the guidebook that was on sale at the exhibition.

The alleged poor resemblance between the panorama and the real thing may have been due to the fact that the panorama showed Hobart as it had been either eight or eleven years earlier. But Darwin seems to be complaining that the scenery is not as attractive. Since the six paintings are reproduced here, readers familiar with Hobart and its surroundings can judge for themselves whether Earle made too much use of artistic licence.

In his Diary, Darwin now gives a brief description of the inhabitants of Van Diemen's Land:

These six watercolours by Augustus Earle were used as the basis for Burford's panorama of Hobart, exhibited in the Strand in London in 1831. The sketches for these paintings were drawn during Earle's two brief visits to Hobart in 1825 and 1828.

The inhabitants for this year are 13,826: in the whole of Tasmania 36,505.— The Aboriginal blacks are entirely all removed & kept, (in reality as prisoners) in a Promontory, the neck of which is guarded. I believe, it was not possible to avoid this cruel step; although, without doubt, the misconduct of the Whites first led to the Necessity:—

With this enlightened conclusion, Darwin aptly summarises the fate of the Tasmanian Aborigines. His statement about their whereabouts, however, confuses two different aspects of the disgraceful history of the treatment of the Aborigines at the hands of the European settlers.

Apparently realising that his Diary account might not be entirely correct, Darwin consulted Bischoff's *Sketch of the History of Van Diemen's Land* (John Richardson, London, 1832) after he returned to England. When the time came to prepare his Diary for publication as the 1839 *Journal*, Darwin was able to give a much more complete version of what had happened:

All the aborigines have been removed to an island in Bass's Straits, so that Van Diemen's Land enjoys the great advantage of being free from a native population. This most cruel step seems to have been quite unavoidable, as the only means of stopping a fearful succession of robberies, burnings, and murders, committed by the blacks; but which sooner or later must have ended in their utter destruction. I fear there is no doubt that this train of evil and its consequences, originated in the infamous conduct of some of our countrymen. Thirty years is a short period, in which

to have banished the last aboriginal from his native island,—and that island nearly as large as Ireland. I do not know a more striking instance of the comparative rate of increase of a civilised over a savage people.

The correspondence to show the necessity of this step, which took place between the government at home and that of Van Diemen's Land, is very interesting: it is published in an appendix to Bischoff's History of Van Diemen's Land. Although numbers of natives were shot and taken prisoners in the skirmishing which was going on at intervals for several years; nothing seems fully to have impressed them with the idea of our overwhelming power, until the whole island, in 1830, was put under martial law, and by proclamation the whole population desired to assist in one great attempt to secure the entire race.

The plan adopted was nearly similar to that of the great hunting-matches in India: a line reaching across the island was formed, with the intention of driving the natives into a cul-de-sac on Tasman's peninsula. The attempt failed; the natives, having tied up their dogs, stole during one night through the lines. This is far from surprising, when their practised senses, and accustomed manner of crawling after wild animals is considered. I have been assured that they can conceal themselves on almost bare ground, in a manner which until witnessed is scarcely credible. The country is every where scattered over with blackened stumps, and the dusky natives are easily mistaken for these objects.

I have heard of a trial between a party of Englishmen and a native who stood in full view on the side of a bare hill. If the Englishmen closed their eyes for scarcely more than a second, he would squat down, and then they were never able to distinguish the man from the surrounding stumps.

But to return to the hunting-match; the natives understanding this kind of warfare, were terribly alarmed, for they at once perceived the power and numbers of the whites. Shortly afterwards a party of thirteen belonging to two tribes came in; and, conscious of their unprotected condition, delivered themselves up in despair. Subsequently by the intrepid exertions of Mr. Robinson, an active and benevolent man, who fearlessly visited by himself the most hostile of the natives, the whole were induced to act in a similar manner.

They were then removed to Gun Carriage Island, where food and clothes were provided them. I fear from what I heard at Hobart Town, that they are very far from being contented: some even think the race will soon become extinct.

By 1845, when the second edition of the *Journal* was being prepared, Darwin had read P. E. Strzelecki's *Physical Description of New South Wales and Van*

Diemen's Land (Longman *et al.*, London, 1845) which gave a more recent report of the Aborigines' fate:

> *Count Strzelecki states, that "at the epoch of their deportation in 1835, the number of natives amounted to 210. In 1842, that is after the interval of seven years, they mustered only fifty-four individuals; and, while each family of the interior of New South Wales, uncontaminated by contact with the whites, swarms with children, those of Flinders' Island had during eight years, an accession of only fourteen in number!"*

Darwin's 1839 account of the removal of the Aborigines from mainland Tasmania contains all the essential elements of the sorry story.[6] The 'one great attempt to secure the entire race' became known as the Black Line. This military-civilian operation commenced on 4 October 1831, with a more-or-less continuous line of about 3000 whites stretching approximately 140 kilometres from Quamby Bluff in the north-central highland region known as the Western Tiers, to St Patricks Head on the east coast. The aim was for the line to move towards the south-east and, as Darwin says, to drive all remaining Aborigines before it, firstly into Forestier Peninsula, and then into the Tasman Peninsula. After

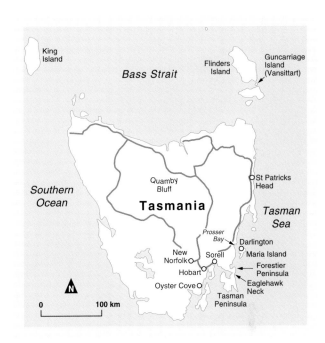

Map of Tasmania, showing places mentioned in the text.

eighteen days of advance, the line was near its goal, stretching a mere 40 kilometres from Sorell to Prosser Bay. However, when scouting parties went ahead to count their captives, they discovered only four Aborigines! The rest, who consisted of only tens rather than the hundreds that were believed to exist, had escaped through the line.

Despite its apparent failure, the Black Line did have an effect on the Aborigines, for as Darwin says, 'they at once perceived the power and numbers of the whites', and consequently 'delivered themselves up in despair'. The government had already appointed George Augustus Robinson as an unofficial ambassador to the Aborigines, and during the next four years, he wandered throughout Van Diemen's Land, convincing the remaining Aborigines that their only hope lay in being brought together in a separate colony, which had already been established on the tiny and inhospitable Guncarriage Island or Vansittart, between Cape Barren Island and Flinders Island in Bass Strait. By January 1832, the settlement had been moved to the larger but still unsuitable Flinders Island, and the number of Aborigines in the settlement stood at 88. The numbers increased over the next few years, as a result of Robinson's activities on the mainland, but the number of deaths, mainly from various European diseases, was alarming.

By 1842, as Strzelecki stated, there were only 54 Aborigines left in the settlement. Five years later, it was decided that the 46 survivors should be resettled at an old convict station at Oyster Cove, just south of Hobart. Neglected by the authorities, and exploited by itinerant sealers and timber workers, they continued to dwindle in numbers until 1876, when the last remaining member of the settlement, Truganini, died. Although she was not the last full-blood Tasmanian Aborigine to die, she was certainly the last of the officially recognised Aborigines, and her death was commonly regarded as representing the extinction of Tasmanian Aborigines as a distinct race. In this regard, Darwin's worst fears about the fate of Tasmanian Aborigines had come to pass.

In fact, as Lyndall Ryan has emphasised in her book *The Aboriginal Tasmanians*,[7] Tasmania still has a substantial population of Aborigines descended from Aboriginal women and the sealers who abducted them. Against substantial odds, these people are continuing their long-standing attempt to establish their rightful role in contemporary Australian society.

Moving on to happier topics, we return to Darwin's Diary, which describes his activities and observations during the next few days:

> 7th ... 10th During these days I took some long pleasant walks examining the Geology of the country:— The climate here is much damper than in New S. Wales & hence the land is much more fertile. Agriculture here flourishes; the cultivated fields looked very well & the Gardens abounded with the most luxuriant vegetables & fruit trees. Some of the Farm houses, situated in retired spots had a very tempting appearance. The

*general aspect of the Vegetation is similar to that of Australia; perhaps it is a little
more green & cheerful & the pasture between the trees, rather more abundant.—*

Darwin's Diary provides no further information about the geological observations that he made during his 'long pleasant walks'. In the 1839 *Journal*, however,
he inserted the following explanation immediately after the first sentence in the
above passage:

> *The main points of interest consist, first in the presence of certain basaltic rocks
> which evidently have flowed as lava; secondly, in some great unstratified masses of
> greenstone; thirdly, in proofs of an exceedingly small rise of the land; fourthly, in
> some ancient fossiliferous strata, probably of the age of the Silurian system of Europe;
> and lastly, in a solitary and superficial patch of yellowish limestone or travertin,
> which contains numerous impressions of leaves of trees and plants, not now existing.
> It is not improbable that this one small quarry, includes the only remaining record of
> the vegetation of Van Diemen's Land during one former epoch.*[8]

This passage indicates the extent of the geological observations made by
Darwin during the *Beagle*'s stay in Hobart, but gives none of the details. For more
information on where Darwin went and what he saw, we need to look elsewhere.
Some clues are provided in the Australian section of his geological book *Volcanic
Islands*, which contains a 1400-word account of the geology of Van Diemen's Land,
together with detailed descriptions of fossil shells and corals (actually Bryozoa)
collected by Darwin. Even better clues are to be found in two manuscripts written
by Darwin during his stay in Hobart and on the voyage home to England.

The earlier of the two manuscripts (the field notes) consists mainly of first
drafts and rough notes. It forms part of the DAR40 papers in the Darwin Archive at
Cambridge University Library, and was transcribed by Dr G. Chancellor from the
Peterborough Museum and Art Gallery and by one of the present authors (JN). The
second manuscript entitled "Memo on Hobart Town", which is part of the DAR38
papers in the same collection, is a secondary account of the geological observations
that Darwin made around Hobart, based mainly on the field notes. Dr M. R. Banks,
formerly Reader in Geology at the University of Tasmania, has published detailed,
annotated transcriptions of these two manuscripts, the latter in collaboration with
Dr David Leaman.[9] The *Volcanic Islands* description of Tasmania's geology is an
abbreviated version of the DAR38 manuscript. The following account of Darwin's
activities in and around Hobart draws heavily on information contained in the two
manuscripts, and on the extensive background information provided by Drs Banks
and Leaman.

On his second full day in Hobart, Darwin wandered around the town, forever on the geological lookout. In the DAR40 manuscript, he records:

Feb 7ʰ — In the town: Sandstones & Greenstones alternately appear, & perhaps in — equal proportion . . . in Government domain — a [Granitic del] Syenitic Greenstone.— Only found one plane of contact,[10] where some sandy shale [ferruginous ins] & Sandstone were in close proximity to a decomposing Greenstone—

Map of Hobart and environs, showing some of the places visited by Darwin.

The following day, Monday 8 February, was devoted to a longer excursion,[11] which was mentioned in the next entry in his Diary:

One of my walks lay on the opposite side of the Bay or river; I crossed in a Steam boat; two of which are constantly plying.— The machinery of one was entirely made here in this Colony; which from its very foundation only numbers 33 years!

The two ferries were the first steamers to operate on the Derwent. One of them, the *Governor Arthur*, was a 63-foot paddle-steamer that was launched from David Callaghan's shipyard at Sullivans Cove on 12 June 1832. In January 1835, she was fitted with a new and larger cylinder that was cast at Captain Wilson's foundry at the lower end of Collins Street. Presumably, this was the local-built machinery that so impressed Darwin. The other steamer was the *Surprise*, which was built in Sydney in 1831.[12]

When Darwin caught the ferry across the river, he landed at Kangaroo Point, near present-day Bellerive. During what must have been quite a strenuous day, he not only collected geological specimens along the foreshore as far as Howrah Point, but he also climbed some of the neighbouring hills (Knopwood and Mornington Hills). The DAR40 manuscript provides some details of what he saw:

> *Walking Coast to South on E side of Ferry: Greenstones; then in close neighbourhead, a quite white, — pottery-like very fine grained sandstone, — in parts <u>rather more Sandy</u> in others frequently more Aluminous or porcelain like (3447:48) . . . passes into & overlies a blueish rock, possessing more of the character of Clay Slate.— These two contain few impressions of Terebratula*

The 'Aluminous or porcelain like' sandstone and the 'blueish' rock were both examples of Ferntree Mudstone; the numbers refer to two samples of the former that Darwin collected and took back to the *Beagle*. 'Terebratula' was the name used by Darwin for all brachiopods (a phylum of simple two-shelled animals). The brachiopod fossils that he saw on this excursion are almost certainly members of the spiriferid order of brachiopods.

After a few more notes, the DAR40 manuscript continues with details of other specimens collected:

> *Ascending the hills, some hundred ft behind the coast, we met with common Greenstone, but the commonest rock is a Greenstone — Syenite [(3453) ins] — On the summit there were gently inclined Strata of altered rocks — Siliceous white, & blue Silico — porcelain rocks (3449:3450) & 3451 . . . On a neighbouring & higher hill, whole rock a dull red ordinary Sandstone.— Do the hard rocks belong to a distinct formation?*

From notes like these, it is evident that Darwin's 'long pleasant walks examining the Geology of the country' were not exactly leisurely strolls through the countryside. If a hill looked interesting, he would climb it. If an adjacent, higher hill looked even more interesting, he would climb it, too. He was always observing or asking questions; always on the lookout for yet another specimen to add to his collection back on board the *Beagle*.

The next morning he was off on another equally demanding excursion, this time along the western shore of the Derwent River, south of Hobart. In the DAR40 manuscript, he records:

> *Tuesday [9th][13] Walked coast South of Town.— Greenstones; generally globular concentric structure:[14] in close connection with inclined strata, (but not inclined by this rock) of Sandstone Conglomerate, containing numerous pebbles of the underlying greenstone; the white Chert with organic impressions &c &c . . . Many of the white pebbles*

strictly resemble the [Porcelain ins] beds of yesterday:— Still travelling onwards,
preponderant quantity of Greenstone

By now he had already walked about 5 kilometres that day – around Battery
Point and Sandy Bay, to the southern end of Long Beach. Then:

— following for a few 100 yards the Basaltic beach, came to cliffs [externally]
composed of cemented fragments . . . This flat point, now formed of such confused,
vesicular rocks probably old Crater

With these words Darwin became the first
person to recognise that the area which stretches
from the southern end of Long Beach to the point
south of Blinking Billy Point is part of the crater
of a long-extinct volcano. To acknowledge Darwin's
insight, in 2008 the point south of Blinking Billy
Point was gazetted as Charles Darwin Cliff.[15]

Moving on, he soon reached and climbed
Porter Hill, where he found many fossils. In the
words of the DAR40 manuscript, they occurred in
'Organic beds' which were

composed of impressions of Betepora & other Coralline
(Mem. Limestone of Argyl) & some few Terebratulae

Polyporella internata, a species
of fossil bryozoan, based on
specimens collected by Darwin
from rocks now assigned to
the Bundella Mudstone on the
shore platform near Porter Hill.
It was described by Lonsdale
(as *Fenestella internata*) in
Volcanic Islands.

In total, Darwin collected nine different
samples of fossils on Porter Hill and around its
base. When they were formally described by G. B.
Sowerby and W. Lonsdale in appendixes to Darwin's
geological book *Volcanic Islands*, six of them were
found to be new species. These were three 'Betepora'
(bryozoans, genus *Fenestella*), two 'Corallines'
(bryozoans, genus *Stenopora*), and one 'Terebratula'
(spiriferid brachiopod).

The next day, Wednesday 10, was unfruitful,
although not through want of trying. Darwin had
decided to climb Mt Wellington, but apparently
wasted most of the day in an unsuccessful attempt. The following day, Thursday 11,
he was successful. In his Diary, he wrote:

11ᵗʰ I ascended Mount Wellington. I made the attempt the day before, but from the thickness of the wood failed.— I took with me this time a guide, but he was a stupid fellow & led me up by the South or wet side. Here the vegetation was very luxuriant & from the number of dead trees & branches, the labor of ascent was almost as great as in T. del Fuego or Chiloe.— It cost us five & a half hours before we reached the summit.—

Drs Banks and Leaman have deduced that on the morning of the 11th, Darwin approached Mt Wellington along the bridle 'road' to Huonville, which left Hobart near present-day Fitzroy Gardens. It is most likely that he walked along the northern side of the valley of the Sandy Bay Rivulet as far as its head, between Turnip Fields and Fern Tree. Near Fern Tree, Darwin collected three different specimens of Fern-tree Mudstone.

During the climb, he recorded observations on the geology of the mountain, as shown in the following passage from the DAR38 manuscript:

Passing over the low ground at its foot composed of the first series, we first reach in the ascent the anomalous flinty and slaty rocks, then come to the Sandstones; these strata extend to a height perhaps of 1200 ft [370 metres], above which there is nothing but Greenstone.— As the Strata on the sides are not very much disturbed, perhaps this height nearly expresses the thickness of this formation.

His Diary description of Mt Wellington continues:

In many parts the Gum trees grew to a great size & the whole composed a most noble forest.— In some of the dampest ravines, tree-ferns flourished in an extraordinary manner:— I saw one which must have been about 25 ft [8 metres] high to the base of the leaves & was in girth exactly 6 feet [1.8 metres]:— the foliage of these trees, forming so many most elegant parasols created a shade approaching to darkness.— The summit is broard & flat & is composed of huge masses of naked Greenstone; its elevation is 3100 ft [950 metres] above the Sea.— [16]

A sample of this greenstone (dolerite) was duly added to his rock collection. The Diary continues:

The day was splendidly clear & we had a most extensive view.— To the Northward the country appeared a mass of wooded mountains of about the same height & rounded outline, as the one we were standing on. To the South the intricate outline of the broken land & water was mapped with clearness before us.— We found a better way to return, but did not reach the Beagle til 8 oclock, after a severe day's work.

Although Darwin does not mention it, his servant Syms Covington also climbed Mt Wellington the same day, presumably with Darwin. We know this from the brief notes that Covington recorded in his own diary. The section describing his ascent of Mt Wellington is as follows:

> Mount Wellington the highest near Town, went up to its summit Feb[y] 11th, from the Town to the Top about 8 or [9] miles, I should suppose but very intricate roads, its top rugged — with low bushes & fresh water in small pools[17]

The next day was Darwin's twenty-seventh birthday, but there is no evidence that he made any fuss about it. In the morning, he went ashore to meet the Surveyor-General. In the Diary, Darwin records:

> 12th ... 15th I had been introduced [to] M[r] Frankland, the Surveyor General, & during these days I was much in his Society:— He took me two very pleasant rides & I passed at his house, the most agreeable evenings since leaving England.

George Frankland had arrived in the colony in July 1827, to take up an appointment as first assistant surveyor, and in March of the next year had become Surveyor-General. By the time of Darwin's visit, the thirty-five-year-old Frankland had been on three exploring expeditions in addition to supervising extensive surveys of land grants to settlers. He was particularly interested in science, having been the foundation vice-president of the Van Diemen's Land Society, which was established in 1830 with the aim of publishing local scientific discoveries as well as establishing a museum and botanical gardens. As one who considered it to be his duty 'to observe and record every remarkable fact connected with the Natural history of the island', he must have been an ideal companion for Darwin.[18]

Darwin's Diary gives no further information on what he and Frankland did during the next few days. But the DAR40 manuscript provides a brief account. On the first day, the entry is:

> Friday 12th.— Called on Surveyor General Mr Frankland.— walked to Lime Kilns & Government Garden =. _Sandstone traversed_ Dined with him in evening.

The 'Lime Kilns' and the associated limestone quarry were located on the western outskirts of the main township, beyond the western end of Burnett Street. The quarry was in the area now bounded by Arthur, Browne, Lochner, and Hamilton Streets. It supplied limestone to Shoobridge's limekiln, which was located in the block bounded by Mary, Burnett, Arthur, and Murray Streets.

In the published account of the geology of Van Diemen's Land in his book _Volcanic Islands_, Darwin describes what he saw:

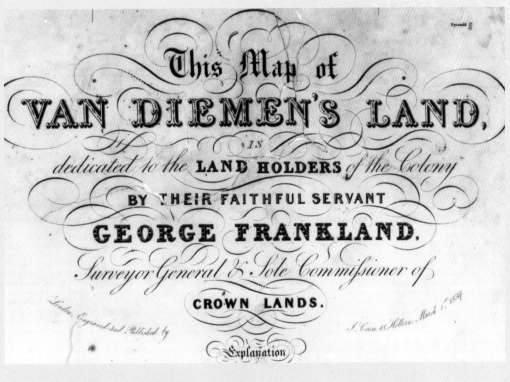

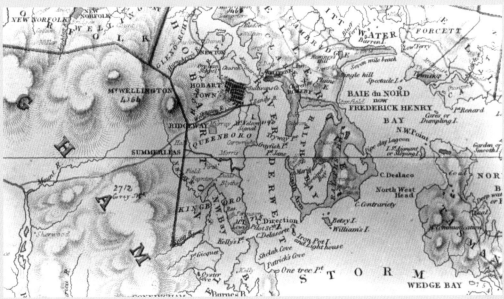

Title and portion of a map showing Hobart and the surrounding areas at the time of Darwin's visit. This map was drawn by George Frankland, the Surveyor-General, who was Darwin's principal host during the *Beagle*'s stay in Hobart.

Behind Hobart Town there is a small quarry of a hard travertin, the lower strata of which abound with distinct impressions of leaves. Mr Robert Brown[19] has had the kindness to look at my specimens, and he informed me that there are four or five kinds, none of which he recognizes as belonging to existing species. The most remarkable leaf is palmate, like that of a fan-palm, and no plant having leaves of this structure has hitherto been discovered in Van Diemen's Land. The other leaves do not resemble the most usual form of the Eucalyptus (of which tribe the existing forests are chiefly composed), nor do they resemble that class of exceptions to the common form of the leaves of the Eucalyptus, which occur in this island.

The travertin containing this remnant of a lost vegetation, is of a pale yellow colour, hard, and in parts even crystalline; but not compact, and is everywhere penetrated by minute, tortuous, cylindrical pores. It contains a very few pebbles of quartz, and occasionally layers of chalcedonic nodules, like those of chert in our Greensand.

From the pureness of this calcareous rock, it has been searched for in other places, but has never been found. From this circumstance, and from the character of the deposit, it was probably formed by a calcareous spring entering a small pool or narrow creek. The strata have subsequently been tilted and fissured; and the surface has been covered by a singular mass, with which, also, a large fissure has been filled up, formed of balls of trap embedded in a mixture of wacke and a white, earthy, alumino-calcareous substance.

Hence it would appear, as if a volcanic eruption had taken place on the borders of the pool, in which the calcareous matter was depositing, and had broken it up and drained it.

This informative report shows that Darwin's visit to the quarry was rewarding. Among other things, it indicates that while examining the excavated area, Darwin discovered some new fossil plants. In addition, Dr Banks has shown that Darwin's observation of the 'large fissure' filled with the mixture of materials is probably the first record of a Neptunian dyke in Australia.

Readers will recall that when Darwin listed the main points of geological interest in Van Diemen's Land in his 1839 *Journal* (see page 131), he stated that this quarry probably 'includes the only remaining record of the vegetation of Van Diemen's Land during one former epoch'. The reason for this claim is clear from the *Volcanic Islands* passage: it arose from the fact that despite extensive searching by local inhabitants, this particular type of limestone had been found nowhere else in the settlement.

As indicated in the DAR40 entry for Friday 12 quoted previously, Darwin walked back to the Government Domain when he had finished examining the quarry, in order to have another look at the sandstone that he had seen there during his second day ashore. In the evening, he dined with George Frankland.

Frankland's house, in which Darwin spent 'the most agreeable evenings since leaving England', had been built in 1831 at Battery Point, on eight acres (3.2 hectares) of land with 1000 feet (305 metres) of waterfront granted to Frankland the previous year. Called Secheron after a fondly remembered site on Lake Geneva, the building still stands today at 21 Secheron Road.[20]

It is not known whether Darwin told Frankland that it was his birthday. If he did, it is most likely that Frankland would have incorporated a small celebration into the dinner that evening.

'I had been introduced [to] Mr Frankland, the Surveyor General . . .
& I passed at his house, the most agreeable evenings since leaving
England.'

Secheron, the house built for George Frankland in 1831, as it
appeared in 1846. This is the house where Darwin dined on the night
of his 27th birthday.

The next day's activities are summarised in the DAR40 manuscript as:

Saturday 13th — Crossed ferry — Rode to examine beds of shells.

This was the first of the 'very pleasant rides' that Darwin took with Frankland. The ferry crossing was the same as the one that Darwin had made five days earlier. But with Frankland as his guide, and with a horse to ride, he was able to travel further once he had reached the other side. In fact, he and Frankland rode to Ralphs Bay or, to be more precise Mortimers Bay, which is about 18 kilometres south-east of Kangaroo Point. As recorded in the DAR40 manuscript, the thing that most interested Darwin was the

> *upraised beach — [small shells ins] Certainly much shells brought by natives — piles stone hatchets — But these may have been upheaved*

Darwin had already noticed what appeared to be raised beaches along both banks of the Derwent River. But the observations he made at Ralphs Bay were used as the centrepiece for his discussion of these beaches in the published geological account in *Volcanic Islands*:

> *Both the eastern and western shores of the bay, in the neighbourhood of Hobart Town, are in most parts covered to the height of thirty feet [9 metres] above the level of high-water mark, with broken shells, mingled with pebbles. The colonists attribute these shells to the aborigines having carried them up for food: undoubtedly, there are many large mounds, as was pointed out to me by Mr. Frankland, which have been thus formed; but I think from the numbers of the shells, from their frequent small size, from the manner in which they are thinly scattered, and from some appearances in the form of the land, that we must attribute the presence of the greater number to a small elevation of the land.*
>
> *On the shore of Ralph Bay (opening into Storm Bay), I observed a continuous beach about fifteen feet [4.5 metres] above high-water mark, clothed with vegetation, and by digging into it, pebbles encrusted with Serpulae were found: along the banks, also, of the river Derwent, I found a bed of broken sea-shells above the surface of the river, and at a point where the water is now much too fresh for sea-shells to live; . . .*

Completing their observations at Ralphs Bay, Darwin and Frankland rode back to the ferry, and returned to Hobart. In the DAR40 manuscript, Darwin records that in the evening he:

> *Dined Attorney General very pleasant musical perf[ormance?] very comfortable [house?] large numerous rooms, beautifully furnished Very pleasant [guest], party.*

The Attorney-General in 1836 was Alfred Stephen, then only thirty-three years old. His house in which the musical party was held still stands at 218 Macquarie Street, on a block of land granted to Stephen in 1825, soon after his arrival in the colony. The concert was in keeping with Stephen's keen interest in music. When describing in his autobiography how he gave up violin practice because his brothers could no longer bear the noise, Stephen joked that 'Thus a Paganini has been lost to the world'. Also, as a young man in London, he recalled that 'Paganini on his unearthly violin [was] a delight to me', and that he went 'often to the theatres, and heard more operas of Rossini and others than I can remember'. In 1839, Stephen left Hobart to take up a judicial appointment in Sydney, where he later became Chief Justice and a leading figure in the colony.[21]

'I dined yesterday at the Attorneys General, where, amongst a small party of his most intimate friends he got up an excellent concert of first rate Italian music. The house large, beautifully furnished; dinner most elegant.'

The Attorney-General in 1836 was Alfred Stephen. His house, which so impressed Darwin, still stands in Macquarie Street. Now known as Stephenville, the house forms part of St Michael's Collegiate School.

After what can justifiably be described as an active week, Darwin had a relatively quiet Sunday. The entry in his DAR40 manuscript is:

Sunday 14th *Staid on board writing letters Stopped from want of* [sun]

Sunday 14 February was a dreadful day in Hobart; it was squally and overcast,[22] there was occasional lightning, and the wind blew at force six or seven for most of the day. Among other things, the stormy weather prevented FitzRoy from taking the sextant readings of the sun that were essential for his determinations of longitude. Consequently, the *Beagle* was unable to depart that day, which accounts for Darwin's comment 'Stopped from want of [sun]'.

Alfred Stephen in 1839, three years after he entertained Darwin to dinner in his house in Macquarie Street in Hobart. This portrait was sketched before Stephen left Hobart to take up a judicial appointment in Sydney, where he later became Chief Justice.

Secure from the inclement weather outside, Darwin settled down in his cabin and reflected on the pleasant time he had had in Hobart. The next entry in his Diary summarises his feelings at that time:

There appears to be a good deal of Society here: I heard of a Fancy Ball, at which 113 were present in costumes! I suspect also the Society is much pleasanter than that of Sydney.— They enjoy an advantage in there being no wealthy Convicts.—

In the 1839 *Journal*, he clarified what he meant in this last sentence by explaining that if there are no wealthy convicts, then there are no

dissensions consequent on the existence of two classes of wealthy residents.

In his Diary, Darwin next concluded that:

If I was obliged to emigrate I certainly should prefer this place: the climate & aspect of the country almost alone would determine me.— The Colony moreover is well governed; in this convict population, there certainly is not more, if not less, crime, than in England.—

With the *Beagle's* departure delayed until the next day, there was time for Darwin to write a letter[23] to his youngest sister, Catherine. After a brief introduction, he informs her that:

> *Tomorrow morning we Sail for King George Sound.— 1800 miles of most Stormy Sea.— Heaven protect & fortify my poor Stomach.—*

He continues:

> *All on board like this place better than Sydney— the uncultivated parts here have the same aspect as there; but from the climate being damper, the Gardens, full of luxuriant vegetables & fine corn fields, delightfully resemble England.—*
>
> *To a person not particularly attached to any particular kind, (such as literary, scientific &c.) of society, & bringing out his family, it is a most admirable place of emigration. With care & a very small capital, he is sure to gain a competence, & may if he likes, die Wealthy.— No doubt in New S. Wales, a man will sooner be possessed of an income of thousands per annum. But I do not think he would be a gainer in comfort.—*

Next, Darwin indicates yet again how strongly he feels about ex-convicts who become rich. Hobart, he tells Catherine, has a much better class of society than Sydney because:

> *Here, there are no Convicts driving in their carriages, & revelling in Wealth.—*

He goes on to say that:

> *You would be astonished to know what pleasant society there is here. I dined yesterday at the Attorneys General, where, amongst a small party of his most intimate friends he got up an excellent concert of first rate Italian Music. The house large, beautifully furnished; dinner most elegant with <u>respectable</u>! (although of course all Convicts) Servants.— A Short time before, they gave a fancy Ball, at which 113 people were present..—*
>
> *At another very pleasant house, where I dined, they told me, at their last dancing party, 96 was the number.— Is not this astonishing in so remote a part of the world?—*

As a young man in his mid-twenties, and especially one who had spent the last four years isolated from the type of social gatherings that he had enjoyed in England, Darwin was agreeably surprised to find so much 'pleasant society' in a settlement as isolated as Hobart. From the passage just quoted, it is evident that the fancy-dress ball mentioned in Darwin's Diary was actually hosted by the Stephens, and that the dancing party was most likely held at Secheron.

The letter continues:

> *It is necessary to leave England, & see distant Colonies, of various nations, to know what wonderful people the English are.— It is rather an interesting feature in our Voyage, seeing so many of the distant English Colonies.— Falkland Island, (the lowest in the scale), 3 parts of Australia: Is^d of France, the Cape.—St Helena, & Ascencion—*

He realises what a marvellous opportunity it is to be able to visit all these places:

> *My reason tells me, I ought to enjoy all this; but I confess I never see a Merchant vessel start for England, without a most dangerous inclination to bolt.— It is a most true & grievous fact, that the last four months appear to me <as> long, as the two previous years, at which ra<te> I have yet to remain out four years longer.— There never was a Ship, so full of home-sick heroes, as the Beagle.—*

He continues:

> *We ought all to be ashamed of ourselves: What is five years, compared to the Soldier's & Civilian's, whom I most heartily pity, life in India?— If a person is obliged to leave friends & country, he had much better come out to these countries & turn farmer. He will not then return home, on half pay, & with a pallid face.— Several of our Officers are seriously considering the all important subject, which sounds from one end of the Colony to the other, of Wool.*

Finally, he was pleased to be able to report that he had not been obliged to draw any money at Hobart:

> *My Father will be glad to hear, that my prophetic warning in my last letter, has turned out false.— Not making any expedition, I have not required any money.—*
>
> *Give my love to my dear Father I often think of his kindness to me in allowing me to come [on] this voyage—indeed, in what part of my life can I think otherwise.—*
>
> *Good bye my dear Katty. I have nothing worth writing about, as you may see,— Thank Heaven, it is an unquestioned fact that months weeks & days will pass away, although they may travel like most arrant Sluggards. If we all live, we shall meet in Autumn.*

> *Your affectionate Brother*
> *Charles Darwin.—*

The next morning, Monday 15 February, the wind had dropped but it was still very overcast and gloomy, and for a second time the *Beagle's* departure was delayed. As Darwin recorded in the DAR40 manuscript, he had a visitor:

> *Monday 15th Mr Frankland came on board, went out riding with him — stormy day — dined in evening most pleasant*

It seems that Frankland was as keen to show his new friend more of the Hobart environs, as Darwin was to be shown. This time they rode north-west of the town, stopping first at a dolerite dyke that occurs between Clare Street Oval and Waverley Avenue, and which is cut by Augusta Road. Continuing further north and west, out past New Town, Frankland showed Darwin a limestone quarry at Barossa Road on the northern flanks of Mt Wellington. In the DAR38 manuscript, Darwin records:

> *Beyond New Town I found a compact crystalline, blackish brown stone, containing some Terebratula, and a few parts almost composed of a small Oyster; there were the impressions of a Pecten; the curious figures, representing Corallines, are chiefly found in the flinty beds.— The Limestone in parts becomes slaty and impure; it is very remarkable by containing an irregular stratum of unequal thickness of Snow white, soft (so as to be excavated with spade and pick) pure Calcareous substance; it is a soft chalk; it is not a little remarkable that such a substance should be included between strata of hard crystalline Limestone.—*

Darwin collected four different specimens of fossils and one of the soft chalk from this quarry. As reported by G. B. Sowerby in his appendix to *Volcanic Islands*, two of the fossils were new species of brachiopods ('Terebratula') which Sowerby named *Producta brachythaerus* and *Spirifera subradiata*.

It is not known if Darwin and Frankland rode further that day. In the DAR38 manuscript, Darwin mentions that:

> *At three or four points, several miles distant from each other, a considerable thickness of the same Limestone strata have been exposed by Quarrys.*

Dr Banks lists possible sites as being Tolosa Street, Collinsvale, and Granton. However, we do not know whether Darwin took Frankland's word for this, or whether he went to see for himself.

After a successful day, Darwin and Frankland returned to Hobart and, as Darwin described in his DAR40 entry for Monday 15 quoted previously, he was Frankland's guest at dinner that evening.

Back on board the *Beagle* after dinner, Darwin wrote a letter[24] to his cousin and fellow beetle-collector William Fox.

After an introduction in which he describes his great disappointment at receiving no letters, he explains that ever since the *Beagle* completed its surveys of South America, its sole remaining task has been to complete the chain of longitude determinations around the globe:

> *Now that the object of our voyage is reduced simply to Chronometrical Measurements, a large portion of our time is spent in making passages.— This is to me, so much existence obliterated from the page of life.— I hate every wave of the ocean, with a fervor, which you, who have only seen the green waters of the shore, can never understand.*

Apart from his apparently never-ending problem with sea-sickness, Darwin hated the long periods at sea because he felt imprisoned. Days at sea represented time wasted: they prevented him from getting on with the things that he wished to do. He continues:

> *It appears to me, I am not singular in this hatred.— I believe there are very few contented Sailors.— They are caught young & broken in before they have reached years of discretion. Those who are employed, sigh after the delights of the shore, & those on shore, complain they are forgotten & overlooked: All think themselves hardly used, that they are not sooner promoted, I thank my good stars I was not born a Sailor.—*

With these thoughts in mind, he has determined never again to go to sea:

> *I will take good care no one shall shall ever persuade me again to volunteer as Philosopher (my accustomed title) even to a line of Battle Ship.— Not but what I am very glad I have come on the expedition; but only that I am still gladder it is drawing to a close.—*

Next he adds a brief summary of his recent scientific activities, in a manner that is even more understated than that used in the letter he wrote to Henslow from Sydney:

> *I have had little opportunity, for some time past of doing anything in Natural History.— I draw up very imperfect sketches of the Geology of all the places, to which we pay flying visits; but they cannot be of much use. Leaving America, all connected & therefore interesting, series of observations have come to an end.—*

This last sentence is the key to why he felt that he had achieved so little scientifically in Australia – he was no longer able to make 'connected & therefore interesting' observations. In contrast, during the first three years and nine months of the voyage, when the *Beagle* was surveying coastlines, Darwin had been able to make many 'connected' observations, because the *Beagle* was continually touching at points up and down both sides of the South American coast, and Darwin was given many opportunities to spend time ashore.

It is worth noting that his 'imperfect sketches of the Geology' of the places visited in Australia are far from being as useless as he suggested in the letter to Fox. In Banks's and Leaman's detailed study of Darwin's observations in Hobart, it is clearly shown that Darwin was a shrewd geological observer. In particular, he was not content simply to describe what he saw: he wanted to know how each set of observations could be related to other observations. By carefully comparing one stratum with another, he was able to deduce the sequence in which various rock formations had been formed. By comparing the types of fossils found in various strata, he was able to draw conclusions as to the relative age of different strata, and to show the relationship between structures at different locations. In this way, Darwin added much to what was previously known about the geology of Tasmania.

Continuing his letter to Fox, Darwin contemplates life after the voyage:

> *I look forward with a comical mixture of dread & satisfaction to the amount of work, which remains for me in England. I suppose my chief <place> of residence will at first be Cambridge & then London.— The latter, I fear, will in every respect turn out most convenient. I grieve to think of it; for a good walk in the true country is the greatest delight, which I can imagine.—*
>
> *I shall find the different societies of the greatest use; judging from occasional glimpses of their periodical reports &c, there appears to be a rapidly growing zeal for Nat: Hist.— F. Hope informs me, he has put my name down as a member of the Entomological Soc:— I do not know, whether you are one.— Formerly, when collecting at Cambridge, how very useful such a central Society would have been to us Beetle Capturers. The banks of the Cam, the Willow trees, Panagaeus Crux Major & Badister, which was not cephalotes, all form parts of one picture in my mind. To this day, Panagaeus is to me a sacred genus.— I look at the Orange Cross, as the emblem of Entomological Knighthood. At Sydney I took a fine species, & long did I look at it, as compared to any other insect.—*

This passage indicates just how keen a beetle collector Darwin was. *Panagaeus cruxmajor* is a small species of carab beetle (family Carabidae) that occurs infrequently in localised areas of England. Only 7 to 8 mm long, it has a black body with striking red markings on each wing cover. *Badister* is another genus of carab beetle, and *Cephalotes* was at that time yet another carab beetle genus.[25] The last sentence just quoted implies that Darwin caught a member of the *Panagaeus* genus, or at least a member of the carab beetle family, while in Sydney. However, none of the published reports arising from his Sydney collections includes any mention of a carab beetle. In any event, most Australian carab beetles lack colourful markings. It is likely, therefore, that he was referring to another type of beetle, possibly *Novius bellus*, a red ladybird beetle with black markings.[26]

After a lament about the length of time before he will next receive any letters from England, Darwin imagines how he will feel when the voyage ends:

> *I think it will be on a September night when we shall first make the Lizard lights. On such an occasion I feel it will be quite necessary to commit some act of uncommon folly & extravagance. School boys are quite right in breaking the binding of their books at the end of the half year & likewise Man of Wars men, when they throw guineas into the sea or light their tobacco pipes with Pound notes, to testify their joy.—*

Then, after giving Fox an account of Australia similar to that given to his sisters, he ends the letter with a romantic description of what he imagines it will be like to visit Mauritius:

> *After touching at King Georges Sound we proceed to the Isle of France.— It will clearly be necessary to procure a small stock of sentiment on the occasion; Imagine what a fine opportunity for writing love letters.— Oh that I had a sweet Virginia to send an inspired Epistle to.— A person not in love will have no right to wander amongst the glowing bewitching scenes.—*
>
> *I am writing most glorious nonsense, so that I had better wish you good night, although at this present moment you probably are just awaking on a cold frosty morning. We are on opposite sides of the World & everything is topsy turvy: but I thank Heaven, my memory is in its right place & I can bring close to me, the faces of many of my friends.*
>
> *Farewell, my dear Fox, till that day arrives, when we shall really once again shake hands. God bless you.—*
>
> *Your affectionate friend*
> *Chas. Darwin*

When Darwin awoke the next morning, it was still too cloudy for FitzRoy to take his sextant readings. In the DAR40 manuscript, Darwin summarises the day's activities as:

Tuesday 16th — Went in coach to New Norfolk — banks of the Derwent — farms — very nice country; mountainous — woody = — walked from town & fell asleep beneath a tree.—

What would have happened if Darwin had slept for so long that he missed the last coach back to Hobart? With the *Beagle*'s departure already so much delayed, would FitzRoy have been willing to delay sailing for another day, just to wait for Darwin to return? Although the answer to this question is almost certainly yes, it is still interesting to contemplate what might have happened if Darwin had been left stranded in Hobart. No doubt, Frankland would have seen that Darwin was usefully employed for as long as it took for the next ship to depart for England. At least his waiting time would not have been as long, nor his circumstances as bad, as those of his former shipmate Augustus Earle when he was left stranded on the island of Tristan da Cunha.

When the notes were translated into his Diary, Darwin mentioned everything except falling asleep:

16th The weather has been cloudy which has prolonged our stay beyond what was expected.— I went this day in a Stage Coach, to New Norfolk. This flourishing village contains 1822 inhabitants. It is distant 22 miles from Hobart town; the line of road follows the Derwent.— We passed very many nice farms & much Corn land. Returned in the evening, by the same Coach.

Once again, the Diary gives few clues about what Darwin observed during the trip. As was the case for the other days, we must look to the DAR40 manuscript to fill in the gaps. Following on from the notes of the previous day, in which he discussed the rocks seen at the limestone quarry in Barossa Road, Darwin wrote:

on road to New Norfolk — we meet the very same stone — This formation much stratified — & strata not very much tilted, only very little.— Perhaps SW? — (Whole town & valley of Derwent, newer formation) — old sst, alternate with much Trappean rock — At New Norfolk, white rocks, similar to those of Kangaroo Point — some more brittle, coarse & siliceous [illeg] Some few pebbles, all siliceous & greater number <u>pure</u> *[<u>white</u> ins]* <u>quartz</u>:

During the walk in which he 'fell asleep beneath a tree', Darwin collected a sample of the 'brittle, coarse & siliceous' rocks.

Although there was a strong north-westerly wind blowing, the day had not been stormy. Consequently, FitzRoy had at last been able to take his sextant readings.

When Darwin arrived back on board that evening, he was told that the *Beagle* was finally ready to sail the next morning.

Before proceeding to the next Diary entry which records the *Beagle*'s departure, we must consider the other available evidence concerning Darwin's observations in Hobart.

Firstly, it is evident from the DAR38 and DAR40 manuscripts, and from the relevant sections of *Volcanic Islands*, that Frankland gave Darwin several samples of rocks and fossils from other parts of Tasmania. Darwin was particularly grateful to receive these specimens, because they greatly extended the information he had collected on comparative relationships. For example, in the DAR38 manuscript, he records:

TOWNS and TOWNSHIPS which are built upon, are printed in Capitals.
TOWNSHIPS not yet settled, in Small Capitals.
Post Stations, Agricultural Settlements, and *Villages,* in Italics.

From Hobart Town to New Norfolk—22 *Miles.*

Upon leaving the town, a little to the right after passing the first mile stone, is a handsome edifice belonging to W. Wilson, esq., J.P. Left, J.H. Emmett, esq. A little farther, on the right, is the residence of Josiah Spode, esq , J.P. At 2 miles left, is the Government Farm; also, the present Orphan School for male children. Roseway Lodge, the seat of H. Nicholls, esq. J.P. Half a mile farther, the new Orphan School and Church. — *2 miles left*

Cross the New-town rivulet, the Rose Inn. There is a brewery and mill to the right. — *3 do., left*

The village of *New-town.*—New-town racecourse. Seats of J. Hone, esq., Captain Swanston, G. Gatehouse, esq., W. Fletcher, esq., John Bell, esq., John Beamont, esq., and others. One mile farther is Derwent Park, the residence of J.T. Gellibrand, esq. — *3 do., right*

Tolosa, the residence of G. Hull, esq. Right, Messrs. Bryants. — *4½ d ., left*

O'Brien's Bridge.—Right, Mr. Branscomb's farm. Mr. Johnson's house and tannery. Cross the rivulet, and enter upon Glenarchy. To the right, a valuable estate belonging to Captain Robertson. — *5 miles*

Green Man Inn. — *7 do., left*

The residence of Mrs. Strickland. — *7¼ do., left*

A farm belonging to Mr. Bilton. A little to the left are also the properties of Mr. Cleburn, Mr. T. Y. Lowes, and others. — *7½ do., left*

Austin's Inn. — *8½ do., left*

Roseneath Ferry, one of the great points of connection between the two sides of the Derwent, and forming the high road to Launceston. — *8¾ do., right*

Stony Point, or Cove Point Ferry, another principal communication between the two sides of the Derwent, and afterwards rejoining the Launceston road. — *9¼ miles*

At Stony Point Ferry, the William the Fourth Inn. A short distance farther, *Hestercombe,* (a post station) at which is Govett's new Ferry, having the same points of communication as the former. — 10 miles

The Hestercombe Inn at the Ferry. — 10 miles

Mr. Govett's farm. — 10 miles

Fox Inn, the principal change house between Hobart Town and New Norfolk. A coach leaves every morning at 8 for the former place, and returns in the evening. — 10½ do., left

Black Snake Inn. — 11 do., left

BRIDGEWATER.—At this place the causeway referred to in page 137, is being constructed across the Derwent. — 11¼ miles

A farm belonging to Mr. Geiss. — 14 do., left

The William the Fourth Inn. An extensive and valuable property adjoins, now in the possession of Messrs. Betts and Co. of Hobart Town. — 15 do., right

Mail Coach Inn. — 16 do., left.

An establishment of lime-burners, rented of Government by Messrs. Betts and Co. On the other side the Derwent, the Lawn, the residence of Arthur Davies, esq. J.P. Fine views of the Derwent all along the road. — 17 miles

Cross the Sorell rivulet ; farms to the left along its banks, belonging to Mr. Williams, Mr. Doran, Mr. Count, and others. — 19 do.

NEW NORFOLK.—Left, quarter of a mile, Mr. Sharland's. Right, Mr. Terry's mill. Cross a bridge over the Thames rivulet, and in a lawn to the right is the Lieutenant Governor's cottage. Right, upon entering the town, E. Dumaresq, esq., the Police Magistrate of the district. Left, the Parsonage and the Invalid Hospital ; also, Murray Hall, the residence of A. Murray, esq A little off, towards the right, is the Bush Inn, Mr. Baker, well known throughout the Colony for its beautiful situation and the excellence of its accommodations. In the same direction are Mason's Hotel, Captain Armstrong's house, Miss Ring's boarding school for young ladies, and the residence of Dr. Officer. New Norfolk has besides, several sub- — 22 do.

Mr Frankland, the Surveyor-General, gave me specimens of the white Aluminous stone, abounding with impressions of Shells from the Huon River and likewise, a blackish Limestone almost composed of parts of Bivalves from the island of Maria.

In a footnote, he adds:

I have also Terebratula from the neck of the Peninsula (where the gards [sic] is kept) of the penal Settlement (3630:31:32).

The main text continues:

We thus see this formation extends over the whole SE extremity of Van Diemen's land.

stantial dwellings, particularly Mr. Turnbull's, Mr. Thompson's, Mr. Joseph's, and others of the same description. Besides the Bush Inn, and Mason's Hotel, New Norfolk has the Star and Garter, an old established and very comfortable house, kept by Mr. Collins, and two other houses on this side the water. as well as four on the other side. At the Bush is a very convenient and well-conducted Ferry establishment, for conveying passengers across the river, and which is particularly recommended to the notice of travellers. A well appointed four-horse coach, runs daily between Hobart Town and New Norfolk.

'I went this day in a Stage Coach, to New Norfolk.'

A contemporary itinerary of the day trip taken by Darwin from Hobart to New Norfolk on 16 February 1836.

The Huon River is about 30 kilometres south-west of Hobart, while Maria Island is off the east coast, about 70 kilometres east-north-east of Hobart. Dr Banks has deduced that the Maria Island specimen is Darlington Limestone from the cliffs and shore platforms north of Darlington. Frankland was particularly familiar with this area, and in fact presented a paper on the geology of Maria Island to the Geological Society on 25 May 1836.[27]

Dr Banks has also shown that the 'neck of the Peninsula' refers to Eaglehawk Neck, the narrow strip of land by which Tasman Peninsula is joined to Forestier Peninsula. At least two of the fossils in these specimens, namely *Spirifera vespertilio* and *Spirifera avicula*, were identified as new species of spiriferid brachiopods in Sowerby's appendix to *Volcanic Islands*. Dr Banks reports that spiriferids are common at Eaglehawk Neck, between Clydes Island and the Blowhole.

Later in the DAR38 manuscript, Darwin states:

I may here mention some facts obligingly communicated to me by Mr Frankland, which will give a general outline of the Geology of the Island.— The central mountains which occupy a large space; and of which Mount Wellington may be considered as the termination in one direction entirely consists of Greenstone.— On their Northern boundary (20 to 30 miles SW of Launceston) there is an extensive formation of Limestone, Conglomerate, and Clay.Slate.— From Quamby Bluff I have specimens of this latter rock marked with impressions of the Corallines and Terebratulate, so frequently mentioned.— Hence there can be little doubt concerning the age of the clay.Slate; and when we consider the variable nature of the Flinty Slate and Limestone beds, containing pebbles near Hobart town; it is highly probable that the whole formation of the North belongs to the same one of which the SE extremity is composed. We shall thus see one continuous series sweeping around the central nucleus of Greenstone.

This passage clearly illustrates how Darwin used his observations to draw conclusions about relationships. The sample of 'Clay.Slate' from Quamby Bluff, which is about 60 kilometres inland from the north coast (and which, incidentally, was the western end of the Black Line described earlier in the section on Aborigines), was not just another sample of fossils; the similarity of those fossils with the fossils that he had 'so frequently mentioned' when describing specimens collected in the south, indicated that the 'Clay.Slate' in the north and in the south was approximately of the same age, and hence of the same formation. While this may seem obvious enough to modern readers, it was not nearly so apparent in 1836. Indeed, Dr Banks has argued that Darwin was amongst the earliest of workers to draw such conclusions about relationships in Tasmania.

We now turn to zoology. Although there is no mention of this topic in the above manuscripts or in his Diary, Darwin (and presumably Covington) in fact caught a number of zoological specimens during the *Beagle*'s stay in Hobart. One of these was an oak skink, which (as it turned out) had not yet been described at that time. When Darwin's specimen was identified by Thomas Bell in part V of the *Zoology*, which was not published until 1843, it was recognised as *Cyclodus casuarinae*, a species that had been first described by Dumeril and Bibron in 1839. Its name was later changed to *Tiliqua casuarinae*.[28] When Darwin captured his specimen, he described it as follows:

> *Scales on the centre of the back light greenish brown, edged on their sides with black; scales on the sides of the body above greyer and with less black, below reddish: belly yellow, with numerous narrow, irregular, waving, transverse lines of black, which are formed by the lower margin of some of the scales being black; head above grey, beneath whitish. . . . It is common in the open woods near Hobart Town in Van Diemen's Land.*[29]

The oak skink collected by Darwin or Covington during the *Beagle*'s stay in Hobart, as illustrated in *Zoology, Part V (Reptiles)*. The modern scientific name is *Tiliqua casuarinae*.

In his zoological diary, Darwin describes five more lizards that he collected in the environs of Hobart. One of them was examined at particularly close quarters:

> *Animal so torpid & sluggish a man may almost tread on it, before it will move.— I lay down close to one & touching its eye with a stick it would more its nictitating membrane & each time turn its head a little further; at last turned its whole body, when upon a blow on its tail ran away at a slow awkward pace, like a thick snake & endeavoring to hide itself in a hole in the rocks.— Appears quite inoffensive & has no idea of biting: held by the tail: collapses its front legs, close to body & posteriorly:— Stomack capacious full of pieces of a white Mushroom & few large inactive Beetles such as Curculios & Heteromeras: Hence partly Herbivorous! — not uncommon on sunny grassy hills:— Tongue, coloured fine dark blue.—*[30]

These last few words remove any doubt as to the identity of the species: it was a blue-tongued lizard – more specifically a blotched blue-tongued lizard, *Tiliqua nigrolutea*. As Darwin noted in his zoological diary, it belongs to the same genus as the oak skink.

In addition to lizards, Darwin collected a snake, which he described as being:

Above coloured "Hair brown with much Liver B[r]" — beneath mottled Grey:—[31]

Of particular interest was his observation that:

The abdomen being burst in catching the animal: a small snake appeared from the disrupted egg: Hence ovoviparous: Is not this curious in Coluber?[32]

'Coluber' refers to the family Colubridae, which at that time encompassed most of the European snakes with which Darwin would have been familiar, namely those that are non-venomous and egg-laying. His question arises because this Tasmanian snake looked like a colubrid and yet it was carrying offspring in 'disrupted' eggs, i.e. in membranous egg sacs that had broken when its abdomen burst. Darwin could see that this snake would be live-bearing rather than egg-laying, and yet it looked as if it belonged to the colubrid family. The answer to his question is that this Tasmanian snake was not a colubrid. It was, in fact, either a tiger snake (*Notechis ater*) or a copperhead (*Austrelaps superba*), both of which are members of the elapid family. In marked contrast to colubrids, elapids are live-bearing and venomous. Apparently, Darwin had no idea that the snake he was examining so closely was actually very dangerous.[33]

Of even greater interest than the lizards or the snakes were the planaria or free-living flatworms. In the zoological diary, Darwin records that during the *Beagle*'s stay in Hobart:

I found beneath a dead rotten tree in the forest a <u>considerable</u> number of this animal.[34]

Later on, he notes that:

I kept some specimens alive in a saucer with rotten wood from Feby 7th to April 1st. When apparently from the excessive heat of the latitude which we then entered they gradually sickened & died.[35]

This passage indicates that Darwin collected some planaria during his second full day in Hobart, and maintained them in his cabin for nearly eight weeks. During this time he studied them intently, noting, for example, their characteristic reaction to light:

I observe they have a particular dislike & immediate apprehension of the light. Directly crawling to the under side of bits of wood.[36]

On 10 February (the day of the unsuccessful attempt to climb Mt Wellington), he bisected one individual in order to investigate planaria's characteristic ability to regenerate a whole individual from part of an individual. Six days later he recorded:

> *On the 16th. Both ends quite lively wounds healing; one orifice manifest in posterior half but none in the anterior.*[37]

In addition to reptiles and planaria, Darwin also collected shells in the environs of Hobart. His list of species includes rock barnacles, *Mesodesma* (a bivalve), a whelk, an amber shell, and some bulimoid land shells, periwinkles, top shells, and air breathers.

Not surprisingly, Darwin also collected insects during his excursions around Hobart, and as in Sydney, his efforts were quite fruitful. Among the Hobart specimens, there were at least 119 species, 63 of which were previously unknown. Included among these new species were dung beetles, leaf beetles, ladybird beetles, weevils, ptinid beetles and parasitic wasps, together with a new water scavenger beetle, a new spider beetle and a new bee.[38]

Darwin found the dung beetles particularly interesting, as they raised important questions concerning adaptation. Towards the end of the 1839 *Journal*, during a discussion of dung beetles found on the Atlantic island of St Helena, Darwin recalled that in Van Diemen's Land he had found:

A gall-forming wasp, *Eucharis iello*, one of the previously unknown insect species collected by Darwin or Covington in the vicinity of Hobart.

> *four species of Onthophagus, two of Aphodius, and one of a third genus, very abundant under the dung of cows; yet these latter animals had then been introduced only thirty-three years. Previously to that time, the Kangaroo and some other small animals were the only quadrupeds; and their dung is of a very different quality from that of their successors introduced by man. In England the greater number of stercovorous [dung] beetles are confined in their appetites; that is, they do not depend indifferently on any quadruped for the means of subsistence. The change, therefore, in habits, which must have taken place in Van Diemen's Land, is the more remarkable.*

This rapid adaptation is still regarded as quite remarkable, especially in view of recent evidence showing that among the *Onthophagus* dung beetles observed by

Darwin, only the Tasmanian members of the species have such catholic tastes in dung: on mainland Australia, three of the same four species are associated solely with the dung of native animals.[39]

The next entry in Darwin's Diary records the *Beagle*'s departure from Hobart:

> *17th The Beagle stood out with a fair wind, on her passage to K. George's Sound. The Gun-room officers gave a passage to England to Mr. Duff of the 21st Reg.*[40]

King George Sound and farewell to Australia

5

AFTER AN UNEVENTFUL VOYAGE ALONG THE SOUTHERN COAST OF AUSTRALIA, LAND NEAR KING GEORGE SOUND[1] WAS FIRST SIGHTED AT 10 O'CLOCK IN THE MORNING OF SUNDAY 6 MARCH. BY 5.30 THAT AFTERNOON, THE BEAGLE HAD TACKED ITS WAY UP THE SOUND AND HAD REACHED THE ENTRANCE TO PRINCESS ROYAL HARBOUR. WITH DAYLIGHT RAPIDLY FAILING, FITZROY DECIDED TO DROP ANCHOR AND WAIT UNTIL FIRST LIGHT THE NEXT DAY.

In his Diary, Darwin noted:

March 6th In the evening, came to an anchor in the mouth of the inner harbor of King Georges Sound. Our passage has been [a] tolerable one; & what is surprising, we had not a single encounter with a gale of wind.— Yet to me from the long Westerly swell the time has passed with no little misery:—

At 5.30 the next morning, the *Beagle* weighed anchor and proceeded through the narrows into Princess Royal Harbour. Two and a half hours later, all sails were shortened and the anchor was dropped in 3 fathoms of water.

Although many navigators, including Captain P. P. King, had visited King George Sound in the preceding decades, it was not until 21 January 1827 that a settlement had been officially established there. Worried that the French might claim possession of the western coast of the continent, and feeling the need for another penal settlement to relieve the pressure on the eastern seaboard, the British Government had instructed Governor Darling to send a party of troops and convicts from Sydney to the sound.[2] Two years later, however, a separate colony answerable directly to London was established on the Swan River, at present-day Perth. By 1831, all the convicts at King George Sound had been transferred to the Swan River, and the settlement had been officially handed over to the authorities at the new colony.[3] Five years later, at the time of the *Beagle*'s visit, King George Sound was showing distinct signs of being a somewhat neglected outpost of the Swan River colony. The contrast with Hobart, which had so impressed Darwin and the rest of the *Beagle*'s company, must have been quite marked.

This panorama of King George Sound was drawn by Lieutenant R. Dale and engraved by R. Howell, for exhibition in London in 1834, just two years before the *Beagle*'s visit. It illustrates many of the features described by Darwin. The viewer is standing on the summit of Mount Clarence. The group of cottages scattered along the edge of the bay is Albany. On the far left, the small clearing with the cottage in the centre is Strawberry Hill Farm, which Darwin and FitzRoy visited.

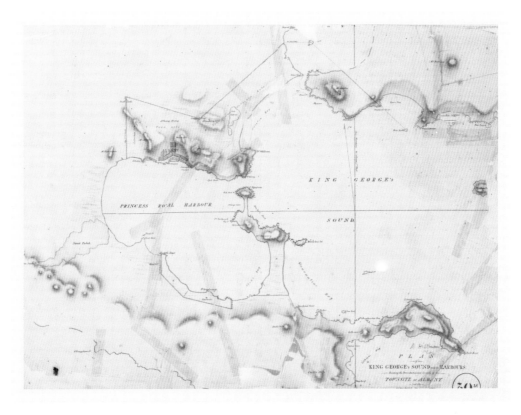

An 1835 map of King George Sound and its harbours. The *Beagle* anchored in Princess Royal Harbour, offshore from the township of Albany. The farm visited by Darwin and FitzRoy is at Strawberry Hill, north-east of the town. Bald Head is at the bottom right-hand corner of the map.

Accordingly, Darwin's impressions, as recorded in his Diary, were not particularly favourable:

> *We staid there eight days, & I do not remember since leaving England, having passed a more dull, uninteresting time.*

FitzRoy's initial reaction was even more unfavourable. In volume II of the *Narrative*, he records:

> *A few straggling houses, ill-placed in an exposed, cheerless situation, were seen by us as we entered the harbour; and had inclination been our guide, instead of duty, I certainly should have felt much disposed to 'put the helm up,' and make all sail away from such an uninviting place.*[4]

Darwin went ashore and climbed either Mt Melville or Mt Clarence, both of which overlooked the infant settlement and the surrounding countryside. In his Diary he wrote:

> *The country viewed from an eminence, appears a woody plain, with here & there a few rounded & partly bare Granitic hills standing up.—*

The Diary continues:

> *One day I went out in hopes of seeing a Kangaroo hunt, & so walked over a good many miles of country.— Every where I found the soil sandy & very poor; it either supported a coarse vegetation of thin low brushwood & wiry grass, or a forest of stunted trees.— The scenery resembled the elevated Sandstone platform of the Blue Mountains: the Casuarina, (a tree which somewhat resembles a Scotch fir,) is however in greater, as the Gum tree is in rather smaller proportion.—*
>
> *There are very great numbers in the open parts, of the grass-tree, these have nearly the aspect of Palm trees, but instead of the crown of noble leaves, there is a tuft like coarse rushes.*

Darwin is here referring to the Western Australian grass tree, *Kingia australis*, or possibly *Xanthorrhoea preissii*. His Diary continues:

The grass-tree, *Kingia australis*, which was named after Captain P. P. King, who collected the type specimen while visiting King George Sound in 1822.

The wiry grass-like plants & brushwood wear a bright green color & to a stranger at a distance would seem to bespeak fertility; a single walk will quite dispel such an illusion; & if he, thinks like me, he will never wish to take another [walk] in so uninviting a country.

The settlement consists from 30–40 small white washed cottages, which are scattered on the side of a bank & along a white sea beach.— There are a very few small gardens; with these exceptions, all the land remains in the state of Nature & the town has an uncomfortable appearance.—

Darwin's description of the settlement is illustrated in a sketch by Syms Covington, apparently drawn from the deck of the *Beagle* at anchor in Princess Royal Harbour. The town certainly does seem to have 'an uncomfortable appearance'.

By far the most encouraging sign in the settlement was the farm belonging to the Government Resident (the official representative of the Governor at Swan River). Darwin's description of his visit to the farm is very brief:

At the distance of a mile, over the hill, Sir R. Spencer has a small nice farm & which is the only cultivated ground in this district.

(DIARY)

Map of Albany and environs, showing the places associated with Darwin's visit.

In volume II of the *Narrative*, FitzRoy fills in some of the details of the visit:

. . . behind a hill, which separates the harbour from the sound, a thick wood was discovered, where there were many trees of considerable size; and in the midst of this wood I found Sir Richard Spencer's house, much resembling a small but comfortable farm-house in England.

This sort of isolated residence has a charm for some minds; but the loss of society, the numerous privations, and the vastly retrograde step necessarily taken in civilized existence by emigrating to perfectly new countries, are I think stronger objections to the plan than usually occur to persons who have not seen its consequences in actual operation.[5]

Captain Richard Spencer had been a half-pay naval officer living in Lyme Regis when he had successfully sought the position of Government Resident at Albany. Armed with a knighthood newly conferred by King William IV (presumably for his many services to the Royal Navy during the Napoleonic Wars), Spencer had

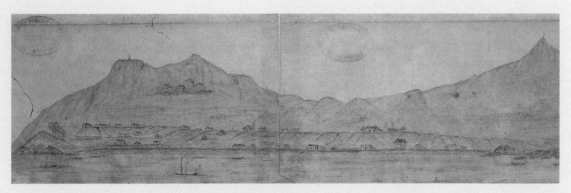

A view of Albany in March 1836, as sketched by Darwin's servant Syms Covington, from the deck of the *Beagle* at anchor in Princess Royal Harbour. Mount Melville is on the left, and Mount Clarence on the right.

On Wednesday, the 9th March, Lady and Miss Spencer laid the first stone in the foundation of each of the new government buildings about to be erected at Albany, under which were deposited medallions bearing excellent likenesses of their present Majesties King William and Queen Adelaide. On the completion of the ceremony, three hearty cheers were given by the spectators, amongst whom were several of the officers of H.M.S. "Beagle."

An article from the *Perth Gazette* of 26 March 1836, reporting an event of some importance that took place three days after the *Beagle* arrived in Albany.

arrived in September 1833, with his wife, nine children, and eleven servants. He was rather surprised to find that the arrival of his party represented a fifty per cent increase in the population of the settlement. On the day that he took up his official duties, Spencer also acquired the lease of the government farm at Strawberry Hill behind Mt Clarence. This area of land had been among the first areas to be cultivated when the new settlement of King George Sound had been founded, and soon became the primary source of food for the inhabitants. Within a few months of his arrival, Spencer had purchased the original farm of 24 acres (9.7 hectares) plus an additional 106$\frac{1}{2}$ acres (43.1 hectares) of uncleared land for what a local historian has described as 'something less than the price of a decent horse'.[6]

'At the distance of a mile over the hill, Sir R. Spencer has a small nice farm.'

The wattle-and-daub cottage visited by Darwin and FitzRoy no longer exists – it burnt down in 1870. It was located in the right foreground of this photograph, abutting the two-storey building, which was built later in 1836, just a few months after the *Beagle*'s visit. The single-storey building to the left was very new at the time of the *Beagle*'s visit – it was constructed in 1835 as a separate kitchen for the cottage. The property, called Old Farm Strawberry Hill, is now open to the public, being owned by the National Trust of Australia (WA).

At the time of Darwin's and FitzRoy's visit, the farmhouse consisted of a small single-storey wattle-and-daub cottage with a thatched roof, which had been built in 1831 by a previous government resident. There was also another cottage to house the servants, a separate kitchen, and sheds and stables. Several months after the *Beagle*'s visit, Spencer commenced building a more substantial two-storey granite house, which still stands today. The property is now owned by the National Trust of Australia (WA), and is open to the public.[7]

Darwin's Diary continues with a catalogue of the shortcomings of the settlement:

> *The inhabitants live on salted meat & of course have no fresh meat or vegetables to sell; they do not even take the trouble to catch the fish with which the bay abounds: [indeed] I cannot make out what they are or intend to do.— I understand & believe it is true, that 30 miles inland there is excellent land for all purposes; this is already granted into allotments & will soon be under cultivation.*
>
> *The settlement of King George's Sound will ultimately be the Sea port of this inland district.— Certainy I have formed a very low opinion of the place; it must however be remembered, that only from two to three years have elapsed since its effectual colonization, & for this, great allowances must be made. Whether however it will ever be able to compete with the Colonies which possess the cheap labor of convicts, time alone will show.—*

As critical as he was, Darwin realised that the settlement did have some positive attributes:

> *They possess here some advantages, the climate is very pleasant, & more rain falls than in the Eastern colonies. I judge of this from the fact, that all the broard flat bottomed valleys, which are covered with the rush-like grasses & brushwood, are in winter so swampy as scarcely to be passable.—*
>
> *The second grand advantage is, the good disposition of the aboriginal blacks; it is not easy to imagine a more truly good natured & good humoured expression than their faces show: Moreover they are quite willing to work & make themselves very useful; in this respect they are very different from those in the other Australian colonies.— In their habits, manners, instruments & general appearance they resemble the natives of New S. Wales.— Like them, they are very remarkable by the extreme slightness of their limbs, especially their legs; yet, without, as it would appear, muscles to move their legs, they will carry a burthen for a longer time than most white men.— Their faces are very ugly, the beard is curly & not at all deficient, the skin of the whole body is very hairy & their persons most abominably filthy.*

Although true Savages, it is impossible not to feel an inclination to like such quiet good-natured men.—

This discussion of the local Aborigines is Darwin's introduction to one of the major events that occurred during the *Beagle*'s visit to Albany. He describes the lead-up to the event in the following terms:

During the first two days after our arrival, there happened to be a large tribe, called the White Coccatoo men, who came from a distance paying the town a visit.— Both these men & the K. George's Sound men, were asked to hold a "Corrobery" or dancing party near one of the Residents' houses. They were tempted with the offer of some tubs of boiled rice & sugar.

FitzRoy also describes the corroboree in volume II of the *Narrative*. He explains that the corroboree was proposed because the local residents wished to 'conciliate' the visiting tribe. He also provides some additional and quite interesting details about who provided the temptations of boiled rice and sugar:

. . . Mr. Darwin ensured the compliance of all the savages by providing an immense mess of boiled rice, with sugar, for their entertainment.[8]

Darwin's account of the corroboree continues:

As soon as it grew dark they lighted small fires & commenced their toilet which consisted in painting themselves in spots & lines, with a white color.— As soon as the dance commenced, large fires were kept blazing, round which the women & children were collected as spectators.—

The Coccatoo & King George's men [formed] two distinct parties & danced generally in answer to each other. The dancing consisted in the whole set running either sideways or in Indian file into an open space & stamping the ground all together & with great force.— These were accompanied each time with a kind of grunt or sigh, & by beating of their clubs & weapons & various gesticulations, such as extending their arms or wriggling their bodies.

It was a most rude barbarous scene & to our ideas without any sort of meaning; but we observed that the women & children watched the whole proceeding with much interest.— Perhaps these dances originally represented some scenes such as wars & victories; there was one called the Emu dance where the set extended one arm in a bent manner: so as to imitate the movements of the neck of a flock of Emus. [In] Another dance a man took off all the motions of a Kangaroo grazing in the woods, whilst another man crawled up & pretended to spear it.— When both tribes

*'these men . . .
were asked to hold
a "Corrobery" or
dancing party.'*

This watercolour
of a corroboree
in the south-
west of Western
Australia was
painted c. 1843 by
Richard Atherton
ffarington. It
illustrates a scene
very similar to
that described by
Darwin.

mingled in one dance, the ground trembled with the heaviness of their steps & the air resounded with their wild crys.—

Every one appeared in high spirits; & the group of nearly naked figures viewed by the light of the blazing fires, all moving in hideous harmony formed a perfect scene of a festival amongst the lowest barbarians.— I imagine from what I have read that similar scenes may be seen amongst the same colored peoples, who inhabit the Southern extremity of Africa.— In T. del Fuego, we have beheld many curious scenes in savage life but never one where the natives were in such high spirits & perfectly at their ease.— After the dancing was over, they formed a great circle on the ground, & the boiled rice, to the delight of all, was distributed to each in succession.—

The next day, Tuesday the 8th, was a busy geological day for Darwin. Following his usual procedure, he made rough notes on the day of the excursion, and then at a later stage prepared a first draft manuscript describing his observations in more detail. Both the rough notes and the manuscript for this and other days at King George Sound have been transcribed and annotated by Dr Patrick Armstrong, School of Earth and Geographical Sciences at the University of Western Australia. Selected excerpts from these transcriptions have been presented in Armstrong's book *Charles Darwin in Western Australia*.[9] The rough notes and the manuscript have also been transcribed by the present authors, and passages from these transcriptions are quoted here.

Darwin's rough geological notes begin:

Tuesday:

Granite on a promontory penetrated by a very great number of veins.— within the space of 80 yards [75 metres] there must have [been] 10 dikes generally 3–4 ft [about a metre] wide. Composed all of one bright green greenstone (3548).

Dr Armstrong has shown that Darwin is here referring to Vancouver Peninsula, which separates King George Sound from Princess Royal Harbour.

When he had finished examining this site, Darwin moved on to Bald Head, at the eastern end of Flinders Peninsula. In his Diary, he recorded:

8th One day I accompanied Capt. FitzRoy to Bald head; this is the spot mentioned by so many navigators, where some have imagined they have seen Coral & others trees petrified, in the position in which they grew.— According [to] our view of the case, the rocks have been formed by the wind heaping up Calcareous sand, which by the percolation of rain water has consolidated & during this process enclosed trees, roots & land shells.— In time the wood would decay & as this took place, lime would be washed into the cylindrical cavities & become hard like stalactites.— The weather is now again wearing away these soft rocks, hence the casts of roots & branches stand out in exact imitation of a dead shrubbery:— The day was to me very interesting, as I had never before heard of such a case.—[10]

As explained by Dr Armstrong, Darwin was correct in attributing the sandy deposits to the action of wind, but he was only partly correct in his explanation of the 'petrified' trees.

The process does begin with the build-up of a calcareous deposit around the plants, but this is not due to the percolation of rainwater through the sand. In fact, it is due to the precipitation of calcium carbonate from underground water that is drawn up by the plants during the regular periods of surface-water shortage that occur in south-western Western Australia.[11] This precipitated calcium carbonate is called calcrete. As the build-up continues, the plants gradually die, and, as described by Darwin, the wood decays and is replaced by the precipitation of additional 'lime' (calcium carbonate). If the wind gradually wears away the material in which this process occurs, then the 'casts of roots & branches' become exposed, and 'stand out in exact imitation of a dead shrubbery'.

The modern term for these 'petrified' trees is rhizoconcretions. As noted by Dr Armstrong, they are not now as frequent on Flinders Peninsula as they were in Darwin's time. However, they can still be seen near Limestone Head, just to the west of Bald Head.

Apart from some notes that indicate a geological excursion on Wednesday 9 March along the northern shores of Princess Royal Harbour, Darwin does not tell us exactly what he did during each of the remaining days in King George Sound. However, from Diary entries cited previously, it is evident that he went on a long walk one day, and from the records of his animal specimens, we know that he was very busy collecting examples of local fauna.

One of his most important finds was a previously unknown species of native Australian rodent: the bush rat, *Rattus fuscipes*, which he 'caught in a trap baited with cheese, amongst the bushes at King George Sound'.[12] This animal is unusual in that it is not a marsupial, but it is a native of Australia. The species inhabits a narrow coastal strip in the south-western corner of Western Australia and much of the eastern seaboard of Australia.[13]

Another interesting find was a southern frog, *Crinia georgiana*, which had first been collected at King George Sound almost two decades earlier by the French naturalists J. R. C. Quoy and J. P. Gaimard. Its present distribution is confined to the south-western corner of Western Australia.[14]

In an earlier section of his Diary, we saw Darwin complaining that the local inhabitants of King George Sound 'do not even take the trouble to catch the fish

Mammalia Pl. 25

The Australian bush rat, a previously unknown species of rodent discovered by Darwin 'amongst the bushes at King George Sound', as illustrated in *Zoology, Part II (Mammalia)*.

The southern frog that was captured by Darwin or Covington at King George Sound, as illustrated in *Zoology, Part V (Reptiles)*.

with which the bay abounds'. Darwin knew that there were numerous fish in the bay because, as FitzRoy recorded in volume II of the *Narrative*, 'During our stay at this place we caught plenty of fish, of twenty different kinds, with a seine'.[15] A few years later, detailed descriptions of ten of these were included in part IV of the *Zoology*. Four of them were claimed to be new species:

Pelates sexlineatus[16]	**Eastern Striped Grunter**,[17] Eastern Striped Trumpeter, Six-lined Trumpeter, Trumpeter Perch
Leviprora inops	**Longhead Flathead**, Crocodile Flathead, Long-headed Flathead, Weed Flathead
Trachurus declivis	**Common Jack Mackerel**, Chows, Cowan Young, Cowanyoung, Greenback Horse Mackerel, Greenback Scad, Horse Mackerel, Jack Mackerel, Scad, Scaly Mackerel
Nelusetta ayraudi	**Ocean Jacket**, Chinaman, Chinaman Leatherjacket, Chinaman Leather-jacket, Chinaman-leatherjacket, Chunks, Leatherjonnie, Leather-jonnie, Yellow Jacket, Yellow Leatherjacket, Yellow Leather-jacket

Subsequently, the first and the last of these were shown to be species previously identified by other naturalists.[18] The other six species in the *Zoology* are:

Arripis georgianus	**Australian Herring**, Bull Herring, Herring, Rough, Roughy, Ruff, Sea Herring, Tommy, Tommy Rough, Tommy Ruff, Western Herring
Pseudocaranx dentex	**Silver Trevally**, Araara, Blue Trevally, Blurter, Bruise-face Trevally, Jack, Ranger, Silver Bream, Silver Fish, Skipjack Trevally, Skippy, Trevally, White Trevally
Aldrichetta forsteri	**Yelloweye Mullet**, Conmuri, Coorang Mullet, Estuary Mullet, Forster's Mullet, Freshwater Mullet, Pilch, Pilchard, Victor Harbour Mullet, Yelloweye, Yellow-eye Mullet, Yellow-eyed Mullet
Pseudorhombus jenynsii	**Smalltooth Flounder**, Jenyn's Flounder, Small Toothed Flounder, Small-tooth Flounder, Small-toothed Flounder
Acanthaluteres spilomelanurus	**Bridled Leatherjacket**, Golden-eyed Leatherjacket, Small Brown Leatherjacket
Family Scorpaenidae	Scorpion Fish.

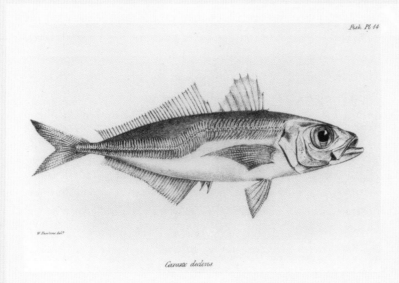

The jack mackerel, one of the previously unknown species of fish discovered by Darwin at King George Sound, as illustrated in *Zoology, Part IV (Fish)*.

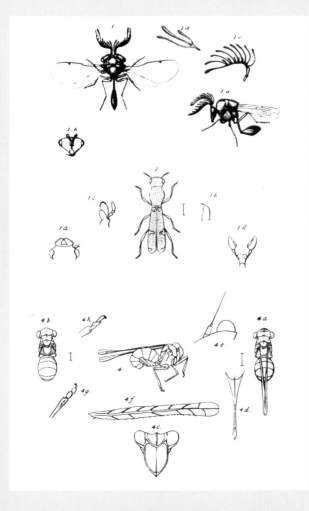

Three previously unknown insect species collected by Darwin or Covington in the environs of King George Sound. Top: a gall-forming wasp, *Eucharis volusus*. Centre: a flower-dwelling beetle, *Allelidea ctenostonoides*. Bottom: an unusual planthopper, *Alleloplasis darwinii*.

Darwin was also active in collecting shells. His zoological diary records that he found an air-breathing limpet, a nerite, a littorinid, a periwinkle, a *Physa* from a freshwater lake (Lake Seppings), and several bulimoid land snails including two species from Bald Head. He was particularly interested in some barnacles that he collected:

> *Balanus. (for dissection) Some Specimens have numerous eggs: or Larva; each of which, when immature is a sharply pointed oval, small animal with six legs, furnished with setae. King Georges Sound. March.*[19]

In addition to collecting the fish and other animals, Darwin also followed his usual custom of collecting insects. During the *Beagle*'s short stay at King George Sound, Darwin (and presumably Covington) collected at least sixty-six species, including forty-eight that were previously unknown. Six of the new species were subsequently named after their discoverer: two planthoppers (*Haplodelphax darwini* and *Alleloplasis darwini*), a predaceous diving beetle (*Hydroporus darwinii*), a small-headed fly (*Ogcodes darwini*; later renamed *Ogcodes basalis*), a seed bug (*Ontiscus darwini*) and a parasitic wasp (*Anipo darwini*; later renamed *Ipoella darwinii*).[20]

Darwin also continued his observations on the planaria he had collected in Hobart. On the day the *Beagle* arrived in King George Sound, the two halves that resulted from his bisection in Hobart were progressing well:

> *Posterior half quite lively; the posterior orifice visible, wound unhealed, crawls in the proper direction: Anterior half with its truncated end quite healed & pointed, slightly pink.*[21]

Unfortunately these two halves were 'lost by neglect' on that same day. Another specimen was then bisected, and observations were continued during the *Beagle*'s stay in King George Sound. In summarising the results of his observations on the regenerative ability of planaria, Darwin later concluded:

> *Thus we see then 25 days sufficed to complete one animal in every respect, and another in its external form & partly in its internal structure.*[22]

Darwin realised that the regenerative and other properties of planaria that he observed were already well known. The significance of his observations is that they were made on a previously unknown species, and thus extended the generality of knowledge about these animals. In a scientific paper published several years after his return to England, he described his Australian species and named it *Planaria tasmaniana*, after the island on which the specimens were collected.[23]

Having spent a week in King George Sound, the *Beagle* was ready to depart. Once again, however, the weather proved to be unfavourable. On Thursday 10 March, for example, the westerly wind was so strong (varying between force six and force seven) that a schooner called the *Sally Ann* was forced to return to port. By the next morning the wind had dropped, but it started to rain, and, as Sir Richard Spencer recorded in his journal, there were 'showers all day'. This turned out to be the beginning of a storm which continued all weekend. On Sunday, Spencer recorded that it was 'Blowing a hurricane' and that there were 'Showers again all day & hail'.[24] By mid-afternoon, however, the worst of the storm had passed, and the *Beagle* at last prepared to depart.

At 5.20 the next morning, Monday 14 March, she weighed anchor and set sail. Twenty minutes later, she ran aground.

The cause of this embarrassing event is not clear. Whatever the reason, the *Beagle* now found herself faced with the prospect of a further delay.

Fortunately, this delay was short-lived. With the aid of a cable and possibly the rising tide, the *Beagle* was set free within one and a half hours, and at last she entered the open sea.

Darwin completed his Diary entry for King George Sound in terms that were not particularly flattering:

> *March 14th Our departure was delayed by strong winds cloudy weather, until this day. Since leaving England I do not think we have visited any one place so very dull & uninteresting as K. George's Sound.*

These comments seem a little at odds with his earlier description of the 'very interesting' visit to Bald Head, and with the number and variety of specimens he had collected. Moreover, they are at odds with the fact that the King George Sound area is located in one of the most distinctive botanical regions in the world – a region noted for the large number of native species (over 5700), and for the exceptionally high proportion (approximately eighty per cent) that are endemic.[25] Furthermore, included among this distinctive flora are some of Australia's most striking plants. Had Darwin arrived in the spring, he would have been most impressed with the beautiful flowers of species such as the scarlet banksia (*Banksia coccinea*) and the showy dryandra (*Dryandra formosa*). But even in March, he could have seen several attractive *Banksia* species and possibly the brilliant red spikes of the swamp bottlebrush (*Beaufortia sparsa*). Had he looked more closely around the swamplands, he would have been fascinated with the insectivorous Albany pitcher plant (*Cephalotus follicularis*), the sole member of the family Cephalotaceae.[26]

However, the last sentence in this passage is not so much Darwin the scientist talking, as Darwin the young man whose thoughts were drifting back to the very exciting and interesting social life he had recently experienced in Hobart. In contrast, the nightlife of Albany in 1836 would have appeared to be 'very dull & uninteresting' to almost any observer.

Besides, it was evident that, for the next few years at least, he would be fully occupied classifying and documenting the huge number of specimens he had collected during the *Beagle*'s voyage. Since this task could be performed only back in England, he could not wait to get home, so that he could commence work in earnest.

Furthermore, the remoteness and isolation of King George Sound would have reminded Darwin that the *Beagle* was still really at the other end of the earth – there was still almost half the globe to be traversed before he would be home again; home to a society in which he would have ready and immediate access to all those aspects of civilisation that were so obviously lacking at King George Sound, and to a lesser extent, in Australia as a whole. He could see that Australia certainly had an assured future, but he did not feel in sympathy with the types of people that were needed to enable the young colony to achieve her potential.

With these thoughts in mind, Darwin says goodbye:

Farewell Australia, you are a rising infant & doubtless some day will reign a great princess in the South. But you are too great & ambitious for affection, yet not great enough for respect; I leave your shores without sorrow or regret.—

(DIARY)

The final comments about Australia in Darwin's Diary.

POSTSCRIPT

in Australia

6

With the *Beagle*'s departure from King George Sound, the main subject matter of this book has come to a close. However, Darwin's involvement with Australia did not end when he left its shores. Indeed, several interesting links between Darwin and Australia were either maintained or established in later years. The aim of this chapter is to summarise these connections, by giving accounts of the subsequent lives of various people who played important roles in the story that has unfolded in the previous chapters.

SYMS COVINGTON[1]

Covington remained with Darwin for two and a half years after the *Beagle*'s voyage, acting as clerk, and helping to organise the huge number of specimens they had collected. At the beginning of 1839, just before Darwin married Emma Wedgwood on 29 January, Covington helped his master move across London from 36 Great Marlborough Street to 12 Upper Gower Street, where the newly-weds established their first home. Once Darwin had settled into married life, it was time for Covington to move on. Apparently impressed with what he had seen of Australia during the *Beagle*'s visit, he decided to emigrate.

Syms Covington.

Darwin was keen to help Covington establish himself in the colony, and towards the end of May 1839, wrote several letters of introduction. One was addressed to the distinguished zoologist William Sharp Macleay, who was a colleague of Darwin's. In March 1839, Macleay had left London for Sydney, in order to help manage the troubled business affairs of his father, Alexander Macleay, who was Colonial Secretary from 1826 to 1837. In his letter to Macleay, Darwin described Covington in the following terms:

The bearer of this letter, Syms Covington, is a young man, who accompanied me, as servant in the voyage of H.M.S. Beagle, round the world.— I have the highest opinion of him and have had means of ascertaining his character. He has saved a little money, and has determined to leave our old world for your new and flourishing one. He means to turn his hand to anything for a few years, and hopes ultimately to become a landowner.— I have perfect confidence in his scrupulous honesty, and as he has been constantly trusted by me with money during the last eight years, I have had opportunities of knowing this.— I should esteem it, a very great personal favour, if you, or any of your family, could give him employment, or put him in the way of obtaining it. I do not hesitate to say, anyone would find him a most useful assistant as a clerk, and I am sure he would soon learn to undertake affairs of considerable trust.— He would, however, much prefer, beginning as a labourer, rather than remain idle.— The only drawback to his advancement is the misfortune of a slight degree of deafness.— During my voyage he shot & prepared nearly all the specimens I brought home, and therefore I venture to hope, that you, who aided me so essentially in publishing their descriptions, will be the more ready to lend him a helping hand, or a little advice (in case he should want it) how to become a good Australian citizen.— You probably would not object assisting (if no other way occurs to you) by a statement that I am a person whose character might be trusted.—[2]

Darwin wrote a second very similar reference for Covington, in the form of an open letter of recommendation;[3] and a third (now lost) was addressed to Captain P. P. King. He also wrote a short letter to the Surveyor-General of NSW, Major T. L. Mitchell, who at that time was on leave in London, and with whom Darwin had recently corresponded in relation to the geology of the Blue Mountains. The letter states:

I am <u>extremely</u> obliged to you, for your great kindness in making enquiries respecting my servant.— But by a great chance, the day after I saw you, I made an arrangement for him to work his passage out as cook to a vessel. I am very sorry you should have taken so much trouble in vain.—

I enclose the curious stone, which I take much shame to myself for not having returned earlier.—but I had stored it away so carefully, that it had utterly past from my mind. I hope before very long, however, to publish a short account of it, & the woodcut, which you permitted to be taken from it.—[4]

The 'curious stone' was a piece of rock that Mitchell had collected during one of his inland expeditions. Darwin did include 'a short account of it, & the woodcut', on pages 38–9 of his second geological book, *Volcanic Islands*, which was published in 1844. He described the stone as a 'volcanic bomb of obsidian', but it is now known to be a flanged-button tektite produced by the impact of a meteorite.[5]

Armed with his letters of introduction, Covington arrived in Sydney in late 1839 or early 1840.[6] Although little is known about his movements during the first few years after his arrival, it seems that the letter of reference to Captain King was the one that led to a job. In April 1839, King had taken up the position of Commissioner (in effect, general manager) of the Australian Agricultural Company, which had extensive pastoral and mining interests in the colony, including cattle and horse studs at Stroud, a settlement about 65 kilometres north of Newcastle. Within a year of his arrival in the colony, Covington married Eliza Twyford from Stroud. Two years later, in 1843, records of the Australian Agricultural Company indicate that Covington was employed as a clerk at the company's coal depot in Sydney.

In that same year, Darwin replied to a letter he had received from Covington, and thus began a correspondence between the two men that lasted until Covington's death. Unfortunately, all of Covington's letters to Darwin, and all but one of Darwin's letters to Covington, have been lost. However, on 9 August 1884, transcriptions of ten of Darwin's letters were published in the *Sydney Mail*.[7]

In addition to telling us a great deal about the relationship between Covington and Darwin, these letters provide a series of personal 'snapshots' of Darwin during the years in which he was formulating and publishing his revolutionary ideas.

The first letter was written a year after Darwin had moved from London to Down House, on the outskirts of the village of Downe[8] in Kent.

The 34-year-old Darwin is now well established as a scientist and author. It is almost five years since he was elected a Fellow of the Royal Society, and he now has twenty-two scientific papers to his credit since returning from the voyage.[9] Also, the first edition of his *Journal*, which was published in 1839, has been very popular with the general public. In addition, the first of the geological books to arise from the *Beagle*'s voyage (*The Structure and Distribution of Coral Reefs*) has been published eighteen months ago, and the last of the nineteen numbers of *The Zoology of the Voyage of H.M.S. Beagle* has just appeared.

> *Down, near Bromley, Kent*
> *October 7th 1843*
> *N.B. This will be my direction for the rest of my life.*

Dear Covington,

Your new ear trumpet has gone by the ship Sultana. It is enclosed in a box from Messrs Smith & Elder to their correspondent, Mr. Evans (I suppose, bookseller). I was not able to get it sent sooner. You must accept it as a present from me.

I presume you will have to pay a trifle for carriage. I recommend you to take your old one to some skilful tinman, and by the aid of an internal plaster cast I have no doubt he could make them.

All that is required is an exact resemblance in form.

I should think it would answer for him to make one, & hang it up in his shop with an advertisement.

I was glad to get your last letter with so good an account of yourself, and that you had made a will. My health is better since I have lived in the country.

I have now three children. I am yet at work with the materials collected during the voyage. My coral-reef little book has been published for a year—the subject on which you copied so much M.S. The Zoology of the voyage of the Beagle is also completed.

I have lately heard that the Beagle has arrived safe & sound in the Thames, but I have heard no news of any of the Officers. Your friends at Shrewsbury often enquire after you. I forget whether I ever [told] you that Mrs. Evans is married & that my father has built them a nice little house to live in.

Captain FitzRoy you will have heard, is gone to New Zealand as Governor. I believe he intended to call at Sydney.

With best wishes for your prosperity, which is sure to follow you if you continue in your old, upright, prudent course.

Believe me, yours very faithfully
C. Darwin.[10]

Among other things, this letter shows that while he was working for Darwin in London, Covington copied the final manuscript for much of *The Structure and Distribution of Coral Reefs.* The publisher of this book, and of the other geological and zoological books that arose from the *Beagle*'s voyage, was Smith Elder and Company. From the above letter, it is evident that Darwin asked Smith Elder to include Covington's ear trumpet in a consignment of books dispatched to a Mr Evans in Sydney.[11]

The second letter was written more than five years later, on 30 March 1849. Darwin's health has deteriorated markedly, but his work still continues. His two other geological books have now been published (*Volcanic Islands*, 1844; *South America*, 1846), as has the second edition (1845) of his *Journal*, which is proving to be even more popular than the first edition. In relating this news, Darwin acknowledges that Covington was the copyist for the first (1839) edition of the *Journal*.

It is now some years since I have heard from you, and I hope you will take the trouble to write me to tell me how you and your family are going on.

I should much like to hear that your worldly circumstances are in good position, and that you are every way fortunate. I hope that your deafness has not increased.

I will now tell you about myself. My poor dear Father, whom you will remember at Shrewsbury, died in his 84th year on 13ᵗʰ November.

My health lately has been very bad, and I thought all this winter that I should not recover. I am now not at home (though I have so dated this letter) but I have come to Malvern for two months to try the cold water cure, and I have already received so much benefit that I really hope my health will be much renovated.

I have finished my three geological volumes on the voyage of the old Beagle, and my journal, which you copied, has come out in a second edition, & has had a very large sale.[12]

Darwin's primary interest at present is in barnacles, and he asks Covington for some Australian specimens:

> I am now employed on a large volume, describing the anatomy and all the species of barnacles from all over the world.
>
> I do not know whether you live by the sea, but if so I should be very glad if you would collect me any that adhere (small & large) to the coast rocks or to shells or to corals thrown up by gales, & send them to me without cleaning out the animals, and taking care of the bases. You will remember that barnacles are conical little shells with a sort of four-valved lid on the top. There are others with [a] long flexible footstalk fixed to floating objects, and sometimes cast on shore. I should be very glad of any specimen, but do not give yourself much trouble about them. If you do send me any, they had better be directed to the Geological Society Somerset House and a letter sent to inform me of them.
>
> I shall not publish my book for 18 months more.[13]

Covington responded to Darwin's request for Australian barnacles, and on 12 March 1850 despatched a box of them to England. Darwin was delighted:

> Dear Covington,
>
> I received your letter of the 12th March on the 25th August, but the box of which you advised me arrived here only yesterday [22 November 1850]. The captain who brought it made no charge and it arrived quite safely. I thank you very sincerely for the great trouble which you must have taken in collecting so many specimens. I have received a vast number of collections from different places but never one so rich from one locality. One of the kinds is most curious. It is a new species of a genus of which only one specimen is known to exist in the world, and it is in the British Museum.
>
> I see that you are one of those very rare few who will work as hard for a friend when several thousand miles apart as when close at hand. There are at least 7 different kinds in the box. The collection must have cost you much time and labour, and I again thank you very sincerely for so kindly obliging me. I have been amused by looking over two old papers you used in packing up, and in seeing the names of Captain Wickham, Mr Maclean and others mentioned. I am always much interested by your letters, and take a very sincere pleasure in hearing how you get on. You have an immense, incalculable advantage in living in a country in which your children are sure to get on if industrious.
>
> I assure that, though I am a rich man, when I think of the future, I very often ardently wish I was settled in one of our colonies, for I have now four sons (seven

children in all and more coming) and what on earth to bring them up to I do not know.[14]

In addition to worrying about his children's future prospects, Darwin was also still very concerned about his own well-being:

I am sorry to say that my health keeps indifferent, and I have given up all hopes of ever being a strong man again.

I am forced to live the life of a hermit, but natural history fills up my time, and I am happy in having an excellent wife and children.[15]

Darwin had commenced his serious study of barnacles (Cirripedia) in 1846, and much of the next eight years was devoted to a detailed investigation of specimens that he obtained from all parts of the globe. In the monographs that were soon to be published describing the results of his extensive study, Darwin described no fewer than 31 species collected in Australia, at locations as scattered as Moreton Bay, Port Stephens, Sydney, Twofold Bay, and Van Diemen's Land. While this may seem to be a rather unconnected set of collection sites, on closer inspection it is evident that each of the three relatively isolated sites (Moreton Bay, Port Stephens, and Twofold Bay) has a distinct connection with one of Darwin's former shipmates. Wickham became police magistrate in Moreton Bay in 1843, and remained there until 1860; P. G. King (Jnr) was manager of the Australian Agricultural Company's horse and cattle studs at Stroud, near Port Stephens, from 1842 to 1851; and Covington settled in Pambula, near Twofold Bay, some time before 1 November 1854, which is the date when he was appointed Pambula's second postmaster.

At a meeting of the Linnean Society of New South Wales held in 1902, Covington's son recalled that as a boy, he had assisted his father 'to collect barnacles for transmission to the author of the *Monograph on the Subclass Cirripedia*; and the Australian species recorded from Twofold Bay in the second volume of this work were derived from this source.'[16]

In his next letter, written on 14 March 1852, Darwin expresses interest in the Australian gold rush, and asks Covington for an account of life on the goldfields. He also gives a brief account of the progress he has made on barnacles:

I have published one book on Barnacles, and am going to publish a second volume, and quite lately I have been examining some of the specimens you sent me, and very useful, and interesting they proved.[17]

Covington did spend some time at the Ovens goldfields in north-eastern Victoria, and on 25 May of the following year, he wrote to Darwin, describing his experiences there. Darwin, responding on 21 October, was disappointed to hear that Covington had not made his fortune out of gold:

> . . . I should have liked to have heard of your turning up a fine nugget worth some hundred pounds, and that would have repaid you for your long journey, which I traced by your letter on the map.[18]

He went on to declare:

> I feel a great interest about Australia, and read every book I can get hold of.[19]

From this time on, Darwin and Covington maintained a regular correspondence, each writing about once a year. The first of this series of letters, written on 28 February 1855, contains little of relevance to the present chapter. Darwin was, however, pleased to hear that things were still going well for Covington:

> I was very glad to get your letter about six weeks ago, dated August 8, 1854, with so good an account of yourself, your affairs, and your children. . . .
>
> How little you thought when we landed together at Sydney, that you should one day have land and house letting £83 per annum. I am very glad to hear that the Colony is progressing so well, and that, as you say, 'our good Queen has no more loyal subjects in her dominion than are the Australians'.[20]

Darwin's next letter, written on 9 March 1856, contains news of his work. All four volumes on barnacles (two on living species, and two on fossil species: comprising a total of 1216 pages of text) have now been published, and in recognition of this work, he has been awarded the Royal Society Medal.[21]

> I was very glad to get a month or six weeks ago your letter of 4th of September, with its interesting account of the state of the Colony and your own affairs, which I am most truly glad are so prosperous. You did a wise thing when you became a colonist. What a much better prospect you have for your sons, bringing them up as farmers—the happiest and most independent career a man can almost have—compared to what they could have been in this old burthened country, with every soul struggling for subsistence. . . .
>
> I have finished my book on the barnacles (in which you so kindly helped me with the valuable Australian specimens). I found out much new and curious about them, and the Royal Soc. gave me their great gold medal (quite a nugget, for it weighs 40 sovereigns), chiefly for my discoveries in regard to these shells, which are not perfect shells, but more allied to crabs.

My health is better, but I have a few bad days almost every fortnight, and cannot walk far or do any hard work. I am now employed on a work on the variation of species, and for this purpose am studying all about our domestic animals and am keeping alive all kinds of domestic pigeons, poultry, ducks. Have you ever noticed any odd breeds of poultry, or pigeon, or duck, imported from China, or Indian, or Pacific Islands? If so, you could not make me a more valuable present than a skin of such. But this, I know, is not at all likely.[22]

Darwin's 'work on the variation of species' was the work to which he returned after finishing with barnacles. As is well known, most of the main ideas of his species theory had been worked out in his own mind nearly two decades earlier, and had been committed to paper as long ago as 1842 and 1844. However, he still needed much more evidence before he would be happy to publish his theory. A major part of this evidence concerned natural variability in domestic animals and plants, and once again Darwin calls on Covington to help him gather specimens.

By the time the next letter was written on 22 February 1857, Darwin had been working solidly on his species book for nine months, but he makes no mention of it in the letter:

I received a short time since your letter of September 14, and was glad to hear how you are getting on, though the account of your affairs was not quite so prosperous as in some former letters, owing, as I understand, chiefly to the expense of your new house. But with your good sense and steadiness I have great hopes that you will ride over the time of difficulty.[23]

Darwin goes on to describe his never-ending concern over the future of his own children. His eldest son is now in his final year at Rugby School:

We think of making him a Barrister, though it is a bad trade.[24]

Darwin still takes a great interest in Australia:

I lately dined with one of your great Australian potentates, Sir W. Macarthur, and heard a great deal of news of Australia, and drank some admirable Australian wine. Yours is a fine country, and your children will see it a very great one.[25]

Covington's new house, to which Darwin refers, and which appears to have been quite costly, still stands in Pambula. It must have been a major venture for Covington, because in addition to providing accommodation for his family, it was also a hotel called the Forest Oak Inn, a post office, and possibly a general store. Now listed on the National Trust's and New South Wales Government's Heritage Registers,[26] the building is presently (2008) a restaurant called Covington's Retreat.[27]

Darwin was late in replying to Covington's next letter, but he had a good excuse: he had been hard at work on his species book. On 18 May 1858 he wrote:

> *I was glad to get some time ago your letter of the 19th August, and I should have answered some time ago, but my health has been very indifferent of late, owing to my working too hard. I have for some years been preparing a work for publication which I commenced 20 years ago, and for which I sometimes find extracts in your handwriting! This work will be my biggest; it treats on the origin of varieties of our domestic animals and plants, and on the origin of species in a state of nature. I have to discuss every branch of natural history, and the work is beyond my strength and tries me sorely.*[28]

Exactly one month after this letter was written, on 18 June 1858, Darwin received a manuscript written by Alfred Russel Wallace, who was then on a natural

The house built by Syms Covington in Pambula on the south coast of
New South Wales.

history expedition in the Malay Archipelago. To Darwin's dismay, Wallace's manuscript contained the essential elements of Darwin's own species theory – the theory of evolution by means of natural selection. Darwin was shattered at the realisation that someone else had independently arrived at the same conclusion as himself, and that furthermore, this other person was now ready to publish, whereas Darwin felt that he still needed to assemble more evidence. Despite his intense disappointment, Darwin offered to arrange for Wallace's paper to be published immediately, and thus to give Wallace sole credit for the theory. However, at the instigation of Charles Lyell and Joseph Hooker, a compromise was reached whereby Wallace's paper was read at a meeting of the Linnean Society held in London on 1 July 1858, together with an extract from Darwin's 1844 manuscript and part of a letter Darwin had sent the previous year to an American colleague Asa Gray. In this way, both men were given credit for development of the idea of evolution occurring by means of natural selection, and at the same time, Darwin's precedence was clearly established.[29]

Despite the supreme importance of this meeting, Darwin was unable to attend because of serious family problems at Down House. In fact, his youngest son Charles had died of scarlet fever just three days before the London meeting, and his eldest daughter Henrietta was seriously ill with diphtheria. After this distressing experience, Darwin took his family away for a break to Sandown on the Isle of Wight.

Now that the general idea of evolution by means of natural selection had been presented to the public, Darwin had to get his species book into print as soon as possible. But by his own reckoning, it would require at least two more years of hard toil before all the evidence was assembled. Clearly he could not wait that long. The solution, urged upon him by his colleagues, was to put aside the work already done on his species book, and as quickly as possible to prepare an 'abstract', outlining his ideas, and summarising the evidence that he had been painstakingly collecting over the last twenty years. According to his personal diary, he commenced the 'abstract' while at Sandown. As many readers know, this so-called abstract turned out to be Darwin's most famous book: *On the Origin of Species by Means of Natural Selection*.

By Christmas Eve 1858, he was more than halfway finished, with 330 manuscript pages written, but his poor health was a great hindrance, and he could work for only a few hours each day. His next letter to Covington, written on 16 January 1859, reflects the severe problems he was facing at this time, but also indicates that, despite the strain, he was still willing to cater to the needs of others:

> *I have got the little book for you, but I have only this moment discovered (for the seal tore by an odd chance at the exact spot) that you asked me to get two copies.*

But I really think that it would be superfluous . . . As to the Aurist,[30] you may rely on it that the man is an advertising humbug. I know plenty of people and have one relation, very deaf, and every one in London would know about this man's power of curing if true. You may depend on it that besides syringing in certain cases there is little or nothing to be done. My father, who was a very wise man, said he had known numbers who had been much injured by Aurists, and none who had been benefited. A common good surgeon can do all that these humbugs can do. I am very sorry to hear about your deafness increasing, it is a very great misfortune for you, but I fear you must look at it as incurable.

I am glad to hear that you are doing pretty well; and if you can settle your sons in an agricultural line they will have no cause to complain, for no life can be more healthy or happy.

We have had an unhappy summer, my eldest girl having been very ill with diphtheria, a new and very fatal throat complaint, and my youngest baby boy having died at the same time of scarlet fever. My second daughter is also very delicate. After our misery we went to Isle of Wight for six weeks for a change. My health keeps very poor, and I never know 24 hours' comfort. I force myself to try and bear this as incurable misfortune. We all have our unhappiness, only some are worse than others. And you have a heavy one in your deafness.

With every good wish for the prosperity of your self and family, believe me, dear Covington, yours very sincerely,

Ch. Darwin[31]

Unfortunately, we cannot follow the aftermath of the publication of *Origin of Species* in this series of letters, because Covington died of 'paralysis' on 19 February 1861, seven months after receiving this latest letter. Only 47 years old, he was buried in the Pambula cemetery, where the headstone records that he was Darwin's assistant on the *Beagle*.

Subsequent events in Darwin's life, including the various reactions to the publication of *Origin of Species*, are described later in this chapter.

CONRAD MARTENS[32]

Having arrived in Sydney seven months before Darwin, Martens remained there for the rest of his long and very productive life. The letter of introduction from Captain FitzRoy to Captain P. P. King succeeded in introducing Martens to the most influential citizens of the colony, many of whom provided him with a more-or-less steady flow of commissions. After initially establishing a studio in Bridge Street, where he was visited by Darwin, Martens moved to Cumberland Street in the Rocks.

In 1837, the year after Darwin's visit, Martens married Jane Brackenbury Carter, the daughter of the colony's first Registrar-General, William Carter. For the next forty-one years until his death, Martens managed to support his wife and their two daughters primarily by the sale of paintings and by taking pupils. His reputation as a skilled landscape artist, especially with watercolours, continued to grow. In August 1847, for example, in William Baker's periodical *Heads of the People*, he was declared to be 'our first [i.e. best] landscape painter'.[33] Despite this recognition, it was not always easy for him to earn a reasonable living, especially during the financially troubled 1840s.

In 1844, Martens had moved with his family to the northern side of the harbour, where he built 'Rockleigh Grange', a house and studio located on five acres of land at the corner of present-day Riley and Edward Streets, North Sydney. Part of the remains of this house is incorporated in the present buildings, which belong to the MacKillop Campus of the Australian Catholic University.

Before settling into 'Rockleigh Grange', Martens had helped the architect James Hume to design the church of St Thomas, which was built between 1843 and 1845 in present-day Church Street, North Sydney, half a kilometre from Martens's house. The Reverend W. B. Clarke, the pioneering geologist, was St Thomas's first rector, and at the church's first vestry meeting, Martens was appointed rector's warden. Between 1881 and 1884, Martens's church was replaced by the larger, present one. However, the sandstone font that he carved in 1845, and the Communion vessels that he donated for the original church, are still in use today.

From time to time, Martens took extensive trips in search of new scenery. In 1841, for example, through his connections with the King family, he travelled north to the Port Stephens area. Another excursion took him to present-day Queensland in 1851.

Eleven years later, and two years after the publication of Darwin's most famous and controversial book, *On the Origin of Species by Means of Natural Selection*, Martens received a message from his former shipmate, via his local rector. Being a keen scientist, the Reverend Clarke had obtained a copy of Darwin's controversial book. Unlike most of his fellow clergymen, both in Australia and elsewhere, Clarke was delighted by it, even writing to Darwin in August 1861[34] with, among other things, extra evidence on glaciation in Australia (which Darwin included in future editions). In his letter, Clarke mentions 'We have living here, an old ship-mate of yours Conrad Martens, who has sometimes named you to me. He is now looking old. He was my first church-warden here in 1846.' In his reply, which was written on 25 October 1861, Darwin thanked Clarke for his support, and added: 'When you next see my old ship-mate Mr Martens pray remember me very kindly to him; I have two of his sketches now hanging up in one of my rooms'.[35]

A sketch by Conrad Martens of the house called Rockleigh Grange that
he built at North Sydney in 1844.

A Martens watercolour showing the original church of St Thomas at
North Sydney.

In response to this greeting, Martens wrote the following letter:

St Leonards
Sydney
Jan 20th/62

To Chas Darwin Esq &c

Many thanks my old Shipmate for your kind message which I have just recd by the padre.

I thought you had quite forgotten that I was in existence, and certainly the man who voluntarily sets himself down in such a place as this has no right to grumble if he [finds] such to be the case.

As it appears however you have still two of my sketches hanging up in your room, I hope you will not refuse to accept another which I shall have much pleasure in preparing and will send to you by the next mail.

Your "book of the season" as the reviewers have it, I must own I have not yet read, altho Mr Clarke offered to lend it me. I am afraid of your eloquence, and I don't want to think I have an origin in common with toads and tadpoles, for if there is anything in human nature that I hate it is a toady, but of course I know nothing of the subject, and they do make such microscopes now a days — I suppose yours is one of Ross's very best, by the by, I got him to make two eyepieces for a reflection telescope just before he died, as I had succeeded in casting and polishing two metals of 6 and 7 feet focus, and so now I show the good people here the mountains in the moon turned upside down, as of course they ought to be when seen from the antipodes.

But I must apologise, for I suppose you don't laugh at nonsense now as you used to do in "Beagle" or rather I suppose it does not come in same way.

Well, that was a jolly cruize, and I hope you have been well and happy ever since — and that you may continue so for some time to come is

Believe me
the sincere wish
of your old shipmate
Conrad Martens

PS

I wonder whether the Admiral "what is now" [is well]. I should like to send my kind regards, if you should see him, but don't if you don't like — coffee without sugar! you remember.—[36]

The painting which Martens sent to Darwin soon after writing this letter is the watercolour entitled *View of Brisbane, 1862*, which was painted from one of the sketches that Martens made during his Queensland excursion. The painting remained at Down House, together with Darwin's other two Martens paintings (the ones he bought in Sydney) for twenty years. After Darwin's death in 1882, Martens's Brisbane painting passed to Darwin's son, who in turn presented it to the Queensland Art Gallery.

The 'Admiral' is of course FitzRoy, and 'coffee without sugar' is a reference to FitzRoy's quick temper, which earned him the nickname of 'Hot Coffee'.

In October 1863, Martens was offered the post of Assistant Parliamentary Librarian at Parliament House in Macquarie Street. The salary of £300 per annum provided him with a degree of financial security, and apparently the responsibilities were not sufficiently great to seriously disrupt his painting. He continued to receive commissions, including one from the Victorian National Gallery in 1872, and another from the trustees of what was to become the Art Gallery of New South Wales, in 1874. This latter commission, a watercolour of Apsley Falls, was the Art Gallery's first acquisition.

View of Brisbane, 1862. This is the painting that Conrad Martens sent to Darwin as a gift early in 1862, after receiving a message from Darwin via the rector of St Thomas, the Reverend W. B. Clarke.

Four years later, in August 1878, Martens died and was buried in St Thomas cemetery (now St Thomas Rest Park, 250 West Street, North Sydney). When the new St Thomas was dedicated in October 1884, Martens's widow donated a second set of Communion vessels, which are also still in use, as a memorial to her late husband.

One of Martens's favourite pupils was Elizabeth Macarthur, the only child of James and Emily Macarthur. Martens's four *Beagle* sketchbooks (the ones mentioned by FitzRoy in his letter of introduction) were bequeathed to Elizabeth, and two of them are still at Camden Park, the Macarthur family home on the outskirts of Sydney. The other two are with the Darwin papers in Cambridge University Library. Many of the works in these four sketchbooks were included in Richard Keynes's *The Beagle Record* (Cambridge University Press, Cambridge, 1979).

AUGUSTUS EARLE

Unlike Martens, Earle lived for only a few years after leaving the *Beagle*. Furthermore, nothing is known of his activities during those years. He died alone in London in December 1838, of 'asthma and debility'. When the *Narrative* was published in 1839, engravings of two of Earle's sketches were included in volume II, which comprised FitzRoy's account of the *Beagle*'s voyage. The first sketch, entitled *Crossing the Line*, shows the traditional ceremonies to which Darwin and other first-timers were subjected during the voyage to South America. The second shows a view of San Salvador, Bahia.

Since a large part of Earle's relatively short working life was spent in and around Australia, he is now regarded as an important Australian artist. An authoritative account of his work and of the many Earle paintings and drawings in the Rex Nan Kivell Collection in the National Library of Australia is given by Jocelyn Hackforth-Jones in her book *Augustus Earle, Travel Artist* (National Library of Australia, Canberra, 1980).

J. C. WICKHAM[37]

The *Beagle*'s second voyage ended officially on 17 November 1836, when she was paid off at Woolwich, after the final chronometer readings had been taken at Greenwich. Less than three months later, she was commissioned for a third surveying voyage. In the meantime, however, FitzRoy had married, and his wife was expecting their first child. In addition, FitzRoy was faced with the somewhat daunting task of preparing both his own journal, and that of Captain King from the first voyage, for publication. As second-in-command of the *Beagle*'s previous voyage, and with FitzRoy obviously otherwise engaged, Wickham was appointed commander for the third voyage.

The voyage eventually commenced on 5 July 1837. Its main task was to explore and survey those parts of the Australian coastline that had not been covered in sufficient detail by the earlier expeditions of Matthew Flinders and P. P. King. The areas of most concern were the north-west coast of Australia, and Torres and Bass Straits.

This was a very different commission from those that had previously sent the *Beagle* to the inhospitable waters of Tierra del Fuego. This time, she would be sailing in relatively calm seas, and she would never be too far from several Australian ports that were already well known to those officers who had been on the previous voyage.

The *Beagle*'s third voyage lasted for six years and three months. During this time, she called in at Sydney on four separate occasions, staying for a total of ten months. As already described in Chapter 3, the officers of the *Beagle* often visited Hannibal Macarthur's family at Vineyard. These visits enabled Wickham to renew his acquaintance with Hannibal's daughter Annie.

Two years after the voyage began, when the *Beagle* was surveying the northern coastline around Port Essington, J. Lort Stokes (second-in-command) and Charles Forsyth (mate), both of whom had sailed on the *Beagle*'s second voyage, took one of the ship's boats to explore a particular opening that had previously been noticed but not investigated. On the morning of 9 September 1839, in the words of the official report of the voyage (written by Stokes), 'A wide bay appearing between two white cliffy heads, and stretching away within to a great distance, presented itself to our view.' On one of the heads they discovered 'a fine-grained sandstone: – a new feature in the geology of this part of the continent, which afforded us an appropriate opportunity of convincing an old shipmate and friend, that he still lived in our memory; and we accordingly named this sheet of water Port Darwin'.[38]

The settlement that was established on Port Darwin in 1869 was initially named Palmerston, after the British Prime Minister who had died four years earlier. In 1911, however, the town's name was changed to Darwin.

Eighteen months after the naming of Port Darwin, the *Beagle* was back in Sydney. Wickham was still seriously troubled by the after-effects of an attack of dysentery he had suffered more than three years earlier, soon after arriving at Swan River. In the end, he decided to transfer his command to Stokes, and to retire from the navy. He sailed for England, but returned to Sydney the next year, and married Annie.

In January 1843, he was appointed police magistrate in the new settlement of Moreton Bay, which later became the city of Brisbane. By the time Wickham's former shipmate Conrad Martens made his only visit to Moreton Bay in November 1851, Wickham and his family were living in a house called 'Newstead', which still stands today.

In the year following Martens's visit, Annie died, leaving two sons and a daughter. In 1857, Wickham was appointed Government Resident in the settlement, and in the same year he married Ellen Deering from nearby Ipswich. When the separate colony of Queensland was about to be formed in 1859, Wickham was offered the post of Colonial Treasurer, but refused the offer on the grounds that he would be left with nothing if he should fail to be elected. By September 1862, he had returned to England, where he renewed contact with his former *Beagle* shipmates Bartholomew Sulivan and Arthur Mellersh. All three visited Darwin at Down House on 21 October 1862.[39]

Wickham then retired to the south of France, where he died on 6 January 1864, at the age of 65.

An interesting footnote to Wickham concerns a Galapagos tortoise known as Harriet, who died in 2006 at Steve Irwin's Australia Zoo, Beerwah, Queensland, having celebrated her 175th birthday the year before. Known as Darwin's tortoise, Harriet was claimed to be one of several young tortoises collected by Darwin during the *Beagle*'s visit to the Galapagos Islands in 1835, and later brought from

A Martens view of Newstead, the house at Moreton Bay (present-day Brisbane) bought by Wickham in 1847 from his brother-in-law Patrick Leslie. This view is from Bulimba, on the eastern side of the Brisbane River, looking towards the west. When Wickham became Government Resident in 1857, Newstead became the unofficial Government House of the settlement.

England to Australia by Wickham at the time he settled in Brisbane. Whilst there is some compelling circumstantial evidence for this claim, a closer examination of documentary and DNA evidence has, unfortunately, shown the claim to be unlikely.[40]

CAPTAIN P. P. KING[41]

In the years following Darwin's visit, King remained active in colonial, business and political life. Along with Alexander Macleay and his son William Sharp Macleay, King was one of the colony's leading scientific figures. He often visited the Macleays at their new mansion, Elizabeth Bay House, which became the scientific centre of Sydney. In addition, he was consulted by most, if not all, of the many visiting naturalists and surveyors who passed through Sydney during the next twenty-five years.

For several months in 1839, King was a Member of the Legislative Council, but resigned after taking charge (as Commissioner) of the powerful Australian Agricultural Company, which at that time employed 500 men (including 400 convicts), and produced 655 bales of wool and 21 200 tons of coal annually. As

Tahlee, the house at Port Stephens (north of Newcastle) occupied by
P. P. King after he became Commissioner of the Australian Agricultural
Company. This sketch was drawn by Conrad Martens in 1841, soon
after King took up his new position.

Commissioner. King moved from Dunheved to a house called 'Tahlee' on the company's extensive estates at Port Stephens, north of Newcastle.

After steering the company through the financial depression of the mid-1840s, King arranged and supervised the sale of large areas of the company's land to individual farmers. By the end of the decade, the company's activities had contracted substantially. In 1849, King travelled to England to confer with the directors about how the financial affairs of the company might be improved. After some negotiations, it was decided that the post of Commissioner would be terminated, and that King would retire.

Freed from these administrative duties, he had more time to devote to the furtherance of science in the colony. Like Captain FitzRoy, he had for many years been interested in recording details of weather patterns. At his Port Stephens home, he had established a private observatory in which he made systematic astronomical and meteorological observations. In addition, barometers and other meteorological instruments were installed at his other houses, and at those of his seven sons. When he was elected as a Member of the Legislative Council in 1851, he installed a barometer in the chamber. Later, he persuaded Sydney University to install a barometer. From the time of Darwin's visit, he had managed to combine his administrative duties with geographical and natural history survey work. He led expeditions to the Murrumbidgee and to New Zealand, and he made extensive surveys of the areas near his two major homes, namely Parramatta and Port Stephens. The results of all these observations – astronomical, meteorological, geographical, zoological, and botanical – were published in various local journals, and in publications produced on his own private printing press.

King also encouraged the Reverend W. B. Clarke in his important geological excursions, especially those in the early 1850s concerned with gold, and edited Clarke's reports. He also maintained a scientific correspondence with Edmund Strzelecki, the Polish count who had spent time on the Australian Agricultural Company's estates at Port Stephens when collecting material for his book *Physical Description of New South Wales and Van Diemen's Land* (Longman, London, 1845). This was the book to which Darwin referred in the second (1845) edition of his *Journal*.

King spent his last years living in a house called 'Grantham', located on the northern side of Sydney harbour, on a site now occupied by the northern approaches to the Sydney Harbour Bridge, very near to the northern end of Ennis Road.[42] Like his fellow parishioner Conrad Martens, he was active in the affairs of St Thomas, North Sydney, becoming a churchwarden in 1854.

Early in 1856, King learnt that he had been promoted to the rank of rear-admiral. A few weeks later, on 26 February, having just returned home after dining on board a ship called HMS *Juno*, then anchored in the harbour, King collapsed and died. His body was ceremoniously transported across the harbour on a barge, to the accompaniment of a twenty-two-gun salute from the *Juno* and another ship called HMS *Herald*. The cortege then proceeded via the newly constructed railway line to Parramatta, and thence to the church of St Mary Magdalene, near his former home at Dunheved. Here the funeral service was conducted by King's fourth son, the Rev. R. L. King, who later became principal of Moore Theological College in Sydney. Standing on land donated by King, and constructed between 1839 and 1840 from bricks made at Dunheved, St Mary Magdalene was very much King's own church, having been named after the church in Launceston, England, where he and Harriet were married. The church still stands today, between King Street and the Great Western Highway at St Marys.

A watercolour by Conrad Martens showing the funeral procession of Rear-Admiral P.P. King, from Milson's Point to Circular Quay. The coffin is being carried on the barge in the centre foreground. The other boats contain officers and crew of HMS *Juno* and HMS *Herald*, which can be seen firing a 22-gun salute in the background. At Fort Macquarie, the Union Jack is at half-mast.

On the following Sunday, a special memorial sermon was preached in St Thomas, North Sydney, by the Reverend Clarke, whose scientific activities had received so much help and encouragement from King.[43]

PHILIP GIDLEY KING (JUNIOR)[44]

After Darwin's departure, King helped his father prepare the manuscript for the first volume of the *Narrative*. At the same time, he became involved with his father's agricultural and business ventures. After spending some time on properties in the Port Phillip district (present-day Melbourne), he surveyed a road in northern New South Wales, and then in 1842 he followed his father and joined the Australian Agricultural Company. Given charge of the company's cattle and horse studs, he

An 1852 watercolour by Conrad Martens of Stroud, where P. G. King
(Jnr) lived from 1842 to 1853.

settled at Stroud, about 30 kilometres inland and north from his father's house at Port Stephens. The next year he married Hannibal and Maria Macarthur's daughter Elizabeth, and thus became Wickham's brother-in-law. By 1851 he was the company's superintendent of stock, and the following year he became assistant superintendent of the company's estates. Soon he had moved to Goonoo Goonoo, a large property on the Peel River about 200 kilometres north-west of Stroud. The property prospered under King's guidance, and a village with its own school and post office grew up around the homestead. Besides managing the property, King was also the manager of the Peel River Land and Mining Company, formed in 1852 by the Australian

An 1861 pencil sketch of Goonoo Goonoo, near Tamworth, where P. G. King (Jnr) lived from 1853 until 1881. The sketch is by P. G. King's son, George.

Agricultural Company, in response to the rush of gold prospectors that followed the discovery of gold in the bed of the Peel River.

While living at Goonoo Goonoo, King corresponded with Darwin. In the earliest extant letter in this correspondence, written on 21 February 1854, Darwin writes:

> *I can hardly tell you how pleased I was, about a week ago, to receive your letter dated the 26th. of October. I lead a rather solitary life, & in my walks very often think over old days in the Beagle, & no days rise pleasanter before me, then [sic] sitting with you on the Booms, running before the trade wind across the Atlantic. Often & often have I wished to hear a little news of you. How changed we are since those days, you with three children, & I with seven, of which the oldest is above 14, & will soon be a young man.—*[45]

He continues with some news of his recent work:

> *I have lately published one volume, & am now preparing a second, on Cirripedes or Barnacles. They have turned out very curious, & were very little known. I have been at the work so many years that I am wearied of the subject: but there is one single species, which I believe is in your Father's collections, namely Scalpellum papillosum of* King *from Patagonia, which I sh^d. like* extremely *to examine, if your Father has a duplicate & if he ever looks over his collections & [cud] lay his hands on it. The species of this genus present a quite new [case] in the animal Kingdom, & have associated with them minute parasitical beings* of the same species; *& which I have called Complemental Males.—*[46]

Most barnacles are hermaphrodites (each individual being both male and female) but Darwin had discovered that some species actually have two separate sexes: the males living as 'minute parasitical beings' in the much larger females. This important discovery of complemental males was described in Darwin's second volume on living barnacles, which was published in the year this letter was written.

After mentioning some news of former shipmates, Darwin concludes:

> *Farewell my dear Philip King, I shall ever think of our old days of friendship with great pleasure; and I hope that your sons may turn out half as nice Boys as you were when you joined the Beagle, & then any parent might be satisfied.*[47]

Eight years later, on 16 September 1862, King wrote to Darwin, asking if his younger brother, who was then on a trip to England, could visit Down House. King also discussed his reaction to *Origin of Species*. He feels that Darwin is probably correct, but he is worried about the implications:

> *My brother John is now on a visit to England & I have asked him if he has time to call on you. . . .*
>
> *I should so much like him to see you that he might bring me a description of, as Jonathan says, Charles Darwin on "location" and of your self and doings. Your work The Origin of Species has a prominent place in my library & was read with much interest. I think you are thought by many to be right who will hardly allow it. I feel in the small scope of my expression that there is much truth in yr deductions, but the question is where do they lead us to — and what is their limit.[48]*

King then continues with a discussion of an issue of considerable interest to Darwin, and to sheep breeders in Australia:

> *I have recently joined in a Newspaper correspondence relative to breeding sheep in the Colony, upholding the idea that we can breed as good sheep as can be imported. I remember a very old letter you once wrote me asking why it was necessary to go on importing. I was too green then to know much about yr queries, but I now ask the same question — One of our most eminent breeders a Mr Bayly has written that Australia is the finest country for producing sheep (ie wool and Mutton especially wool) yet he continued unless occasional importations are made, a degeneracy ensues — with these opinions he coupled another that he has sheep without a fault. I thought his logic so bad that I pressed him for a cause for the degeneracy — but he could not explain at all what it proceeded from. I believe it is caused (for that there is degeneracy no one can doubt) by continuously feeding sheep in large flocks, hunting them with dogs, obliging them to travel every day 4 or 5 miles from food & keeping them 14 hours out of the 24 without it. If we could afford with cheaper labour to keep sheep as they are kept in Germany we might soon export Rams there. If you ever publish your promised large work on yr favourite subject I shall look out for it eagerly.[49]*

King clearly appreciates that degeneracy in physical appearance, due to the relatively harsh conditions in which sheep are raised in Australia, does not in itself cause genetic degeneracy. Noting that this letter was written just before Mendel had completed his pea-breeding experiments, and almost forty years before the mechanisms of inheritance were rediscovered and generally accepted, it is evident that King is showing considerable insight.

The letter ends on a personal note. Realising that they both will have changed over the years, King asks Darwin for a photograph:

> *We have now grown up families, & I think we should never recognise each other. I wish you would send me your carte de Visite if you ever indulge in such matters . . .[50]*

Darwin replied on 16 November, as soon as he received King's letter:

I was much pleased to get your letter; only about a fortnight since I was asking all I could about you, for Wickham, Sulivan & Mellersh came down here; & much I enjoyed seeing these good & true old Beagle friends. I wish you had told me a bit more about yourself & family; for the remembrance of old days when we used to sit & talk on the booms of the Beagle, will always to the day of my death make me glad to hear of your happiness & prosperity.

I have no proper carte, but I send a photograph of myself made 3 or 4 years ago by my eldest son (now a partner in a Bank,— this shows how old I am) & which is a good likeness of me.— Thanks for your information about sheep; I remember being puzzled about their degeneracy.—

But to the main purpose of your letter, I grieve to say that my health is so indifferent, I cannot stand seeing at present anyone here. Twice lately I could not resist seeing old friends (once was when Wickham & Co came here) & the excitement made me so ill afterwards, that I have been advised not to do so again. I am well enough in the mornings & when I keep quiet. I must write to your Brother to this effect. . . .

Do you remember Syms Covington? I had a letter about a year ago from Twofold Bay to say he was dead.[51]

When King received Darwin's letter with the photograph enclosed, he became very nostalgic:

Goonoo Goonoo   NSW
Apr 19 / 63

My dear Mr Darwin
You cannot think with which pleasure I received a note from you by the last Mail enclosing a photograph of yourself. It was a strange feeling that came over me as I identified one by one the now well remembered features of what you must permit me to call yr dear old face. And as I gaze upon your likeness reminiscences of my youth appear to come back to me. I fancy I see you in your old corner behind the mizen-mast with neither room to bend or turn still pursuing your studies in Natural History with unwonted zeal. But I will not bore you with these reminiscences suffice it to say that few things have given me greater pleasure lately than the opportunity you have given me of reproducing in my minds eye the very expression of your eye & mouth.

I am so sorry to think you suffer so much from ill health but you are living on to a good age at any rate. And I trust the Scientific world will not lose you yet. Accept my dear friend the best wishes of yr old shipmate

Philip Gidley King[52]

Several years passed before King next wrote to Darwin:

Sydney
Feb 25 / 69

My dear Mr Darwin,

On reading the enclosed I could not help thinking of you. I suppose the variety is a Sport — but who can say what might not be made of it by crossing & judicious selection — Did you ever get any answers to your queries about the habits and manners of our blackfellows. I tried to make some answers but I found myself unable to distinguish the aboriginal manner from the acquired habit. All blacks I have associated with have been more or less civilized.

I often think of you & read yr works. I obtained the first copy of yr Domesticated Animals etc that arrived in the Colony & noticed with much pleasure yr mention of my name.

I hope you [enjoy] better health than formerly — you & I will soon be the last relics of the old Beagle. Don't trouble to reply to this — [though] I am always pleased to get a line from you. I may take a trip home one of these days & will hunt you up.

[Very] sincerely
Philip Gidley King [53]

King's comments about the habits and manners of Australian Aborigines relate to the many queries that Darwin made in collecting information for his book *The Expression of the Emotions in Man and Animals*, which was published three years later, in 1872. The other book mentioned by King is *The Variation of Animals and Plants under Domestication*, published in January 1868. This book is really the first two chapters of the comprehensive 'species book' that Darwin was forced to set aside in 1858, after the arrival of Wallace's paper on natural selection. It provides the detailed evidence that Darwin omitted from the chapter entitled 'Variation under domestication' in *Origin of Species*. In discussing the domestication of dogs in chapter 1 of the *Variation* book, on page 21 Darwin says, 'Mr. Philip King informs me that he once trained a wild Dingo puppy to drive cattle, and found it very useful.' Later in the same chapter, on page 29, he says, 'Even the wild Dingo, though so anciently naturalised in Australia, "varies considerably in colour," as I am informed by Mr. P. P. King.' It is these passages to which King refers in his letter.

Darwin's reply to King's letter is the last of the known King–Darwin correspondence:

Down.
Beckenham
Kent. S.E.
Ap. 24. 1869

My dear King

    Although you tell me not to write I must send a line for auld lang syne & to thank you for yr notes.

    Living the quiet life which I do I often think of old days, & I remember the friendship of no one on board the Beagle with more pleasure than yours.

    The cases of the parrot seem very curious, but at present I do not know what species the rosella is; I will however when next in London, shew the paragraph to Mr Gould & find it out. Many thanks for your kind wish to aid me about expression.

    I can give you no news about any other of our old shipmates as I have heard of none of them for some months. As for myself my health continues & will ever be weak, but I am better than I was a few years ago. Should you come to England it would be a real pleasure to me to see you.

    Believe me my dear King
    Yours very sincerely
    Charles Darwin[54]

King remained as manager of Goonoo Goonoo for almost thirty years before handing over responsibility to his son in 1881, and retiring to a house called 'Banksia' in the eastern Sydney suburb of Double Bay.

A year after King settled in Sydney, Charles Darwin died. Five months later, King heard from Archibald Liversidge, the Professor of Geology at the University of Sydney, that Darwin's son Francis was collecting material for his *Life and Letters of Charles Darwin*. On 8 September 1882, while on a trip to Adelaide, King wrote to Francis Darwin:

My dear Sir

    You will find my name amongst the officers of the Beagle, and you will then know that I was the shipmate of your most revered father — some years ago he sent his photograph taken as he said by one of his sons, and I have had several notes from him during the last 20 years. Professor Liversidge has just informed me that you are anxious to have copies of any private correspondence and I now write briefly to say

that immediately on my return to Sydney I will copy such as I have and send to you.
You can then say whether there is anything in them you wd like to preserve—

Your dear father always remembered with expressions of affection the "delight-
ful evenings" he and I "used to spend sitting on booms of the Beagle in the Tropics."
I perfectly remember the charm he gave expression to on experiencing the delightful
sensation of tropical airs — wafted [illeg] out of the sails overhead.

My heart came into my mouth when I learnt that England had placed your
dear Fathers remains in Westminster Abbey, a fitting tribute to so great a man.[55]

As soon as he had returned to Sydney, King dispatched a copy of Darwin's letter of 16 November 1862. In the end, Francis Darwin did not include this letter when *Life and Letters* was eventually published five years later, in 1887. However, in the introduction to chapter 6,[56] Francis acknowledges his correspondence with King, and quotes part of the second sentence of his father's letter to King.

In 1890, King received a request from John Murray, the company who had published all of Darwin's later books, and who had also published Francis Darwin's *Life and Letters*, to write an article describing his experiences with Darwin on the *Beagle*.[57] By November 1891, he had made no progress on the article, but he had prepared some 'diagram sketches of the old ship's interior arrangements', and he sent copies of these to Murray. They appeared in subsequent John Murray printings of the 1845 *Journal*.[58] The originals are held in the Mitchell Library in Sydney, and are reproduced in this edition on pp. 12–13.

King was quite busy in his retirement, being appointed to the Legislative Council, and becoming a director of the Mercantile Bank of Sydney and president of the Australia Club. In addition, he wrote a book entitled *Comments on Cook's Log (H.M.S. Endeavour, 1770) with Extracts, Charts and Sketches*, which was published by the New South Wales Government Printer in 1891. By October of the next year, he had made substantial progress on his reminiscences of life on board the *Beagle* with Darwin, and dispatched a five-page manuscript to Murray. This was never published.[59] It does, however, contain some amusing anecdotes, including an account of Darwin's initiation by King Neptune during the ceremony of crossing the line. It also contains the interesting suggestion that one of the factors that encouraged Darwin to sail with the *Beagle* in the first place was that Captain P. P. King was willing to send his own son (P. G. King – the author of the manuscript) on the voyage.

King survived for another ten years before dying in August 1904 at the age of 86. Like his father before him, he was buried in the churchyard of St Mary Magdalene near Dunheved.

CAPTAIN FITZROY[60]

When the *Beagle*'s voyage was completed in October 1836, FitzRoy was only 31 years old. After marrying Maria Henrietta O'Brien, the daughter of a major-general, within months of landing, he set about the task of preparing his own account of the *Beagle*'s voyage. With Captain King settled in Australia, FitzRoy was also given the task of seeing King's account of the first *Beagle* voyage through to print. In the end, as already described in previous chapters of the present book, three volumes of the *Narrative* were published (King, FitzRoy, and Darwin), plus a detailed appendix to FitzRoy's volume, containing comments on all manner of topics relating to the voyage, together with the weather records he had so diligently collected, using Beaufort's wind scale and weather code. Even before the results of the voyage were published, the quality of FitzRoy's navigating and surveying work was recognised by the award in 1837 of the Royal Geographical Society's gold medal.

With the *Narrative* published, and with a family to care for, FitzRoy's life changed direction, and he was elected to parliament in the general election of 1841, as one of the two members for the County of Durham. In the course of the election, FitzRoy became embroiled in a row with another candidate, which proceeded through one fantastic episode after another, culminating in a street brawl between the two men outside the United Services Club in the Mall. Although he was by no means totally to blame, FitzRoy had shown clearly that his notorious temper was still not always under control.

Gravestones of Philip Gidley King (Jnr) and of his wife Elizabeth (daughter of Hannibal Macarthur) in the churchyard of St Mary Magdalene, Great Western Highway, St Marys.

Once in parliament, FitzRoy soon busied himself with naval matters, and introduced a bill to ensure that command of merchant vessels was given only to those men who passed specified examinations on navigation and other practical aspects of seamanship.

After only two years in parliament, he was asked to become Governor of New Zealand. Although in theory this was a prestigious position, in practice it was a thankless task: New Zealand at that time was torn by irreconcilable conflicts among and within the races, and even a skilled and tactful colonial administrator would have been almost certainly doomed to failure. As it was, not even FitzRoy's greatest admirers would have claimed that he was overly endowed with either of these qualities.

Soon after arriving in the colony, FitzRoy wrote to his former commander Captain P. P. King, outlining the problems that he faced:

> *I found things in a most extraordinary state of confusion — and I am so crippled by the injunctions of the Home Government that it is quite bewildering.*
>
> *I shall probably be recalled before long for doing too much — for acting without sufficient authority — or against my instructions — but this I must risk — in such a state are public affairs.*[61]

Confronted with this 'most extraordinary state of confusion', FitzRoy felt that he had no option but to take matters into his own hands – to act 'without sufficient authority'. If his policies had succeeded in creating order out of chaos, the Government might have looked favourably on his actions. As it was, FitzRoy's policies did not work, and the political situation deteriorated. This left the Government with little choice but to relieve him of his commission in less than eighteen months, and to recall him to London.

Soon after arriving home to something less than a hero's welcome, FitzRoy must have been disappointed to learn that his half-brother Charles, who had much better social connections, had been appointed to the far more prestigious, and far less troublesome, position of Governor of New South Wales. Worse still, Charles FitzRoy was later regarded by many people as having done a good job during his term of eight years, and he was duly rewarded with a knighthood.

In the meantime, FitzRoy became involved once again with naval affairs, this time being appointed Superintendent of Woolwich Dockyard, in which capacity he was responsible for conducting trials of the Royal Navy's first screw-driven steamship.

By now, however, he had decided that his talents would be put to better use if he devoted himself to the infant science of meteorology. Accordingly, he resigned from the Navy, and after being elected a Fellow of the Royal Society for his services to navigation and surveying, he was put in charge of the newly established Meteorological Section (later to become the Meteorological Office) of the Board of Trade. He set about this new task with his customary energy and enthusiasm, and was extremely effective at organising the regular collection of weather data from ships' captains and from coastal towns. Before long, he was issuing sturdy barometers (known as FitzRoy barometers) together with clear instructions as to their use, and was utilising the data collected from these instruments to issue storm warnings. Responding to the need to summarise the information he was collecting in a readily understandable format, he developed what he called synoptic charts. FitzRoy then used these charts to produce the world's first daily weather forecasts, which were published in newspapers.

All this rather frantic activity culminated in the publication in 1863 of his *Weather Book: a Manual of Practical Meteorology* (Longman, Green, Longman, Roberts & Green, London), a comprehensive 440-page account of the principles and practice of weather recording and forecasting. Demand was sufficiently strong for a second edition to be published the following year.

Despite this obvious success, a number of matters had caused FitzRoy great concern and distress. His wife had died in 1852, followed two years later by his eldest daughter, then 16 or 17 years old; Sulivan, his junior officer on the *Beagle*, had beaten him to a senior job in the Board of Trade; and his former cabin mate had published a theory of evolution that seemed to leave no room for a literal interpretation of the Bible, which had become increasingly dear to FitzRoy over the years. To make matters worse, it was clear that one of the main stimuli that had led Darwin to reach his heretical conclusion was the information he had collected during the voyage of the *Beagle*. And it was he, FitzRoy, who had invited Darwin to go on the voyage.

FitzRoy's worst fears about Darwin's theory were realised when he chose to present a paper on British storms to the 1860 meeting of the British Association for the Advancement of Science. Unbeknown to FitzRoy, and to everyone else at the time they submitted their papers, this was to be the famed Oxford meeting at which Thomas Huxley and Bishop Wilberforce debated Darwin's theory. Along with just about everyone else who happened to be in Oxford on Saturday 30 June 1860, FitzRoy attended the lecture at which the debate occurred, and he even contributed to it, stating how he regretted the publication of Darwin's book, and explaining how he had often warned his former shipmate not to entertain ideas that were contradictory to the Bible.

During the next few years, FitzRoy gradually deteriorated in both physical and mental health, owing at least in part to his increasing sensitivity to criticism of the inevitable inaccuracies in his weather forecasts. On 30 April 1865 he committed suicide.

Notwithstanding this tragic end, it can be seen that FitzRoy had a number of significant achievements to his credit. Although he is now remembered by most people only as the captain of the ship on which Darwin sailed, it is probably more just to remember him as a talented navigator and surveyor, and as the 'father' of weather forecasting.

CHARLES DARWIN

In considering the subsequent activities of the other people described in this chapter, we have also covered much of Darwin's later life. We have seen how he initially settled in 36 Great Marlborough Street in London, where for the next two and a half years Syms Covington helped him to arrange and document the material that had been collected during the *Beagle*'s voyage. We have followed his move to 12 Upper Gower Street in London, upon his marriage to Emma Wedgwood in January 1839; and his family's final move, nearly three years later, to Down House in Kent.

Mention has also been made of the publications that arose directly from the voyage: the nineteen numbers of *The Zoology of the Voyage of H.M.S. Beagle* (1838– 1843); *Journal of Researches* (1839, 1845); and the three volumes of *The Geology of the Voyage of the Beagle* (*Coral Reefs*, 1842; *Volcanic Islands*, 1844; and *South America*, 1846), followed in 1851 and 1854 by his four monographs on barnacles (Cirripedia).

We have also seen how, during all this publishing activity, he was gradually collecting evidence to support his 'species theory', the essential elements of which had been written down as early as 1842. The profound effect of the arrival of Wallace's paper in June 1858 has also been described, as has Darwin's resultant decision to set aside his species book, and to concentrate instead on an 'abstract' of his theory, which became his most famous book, *On the Origin of Species by Means of Natural Selection* (1859). We have also followed activities at the Oxford meeting held in June 1860, and have witnessed first-hand accounts of the reactions of some of his former shipmates.

The more general Australian reaction to Darwin's theory has been documented by Ann Moyal in *A Bright and Savage Land* (Collins, Sydney, 1986). Briefly, the initial reaction was quite hostile, even from scientific colleagues such as W. S. Macleay. Ironically, one of the few Australian-based people who actively supported Darwin in the years immediately following publication of the *Origin* was W. B. Clarke, who, as we have already seen, was a clergyman. The reason for this unusual state of affairs may lie partly in the fact that both Clarke and Darwin had very similar educations. Both went up to Cambridge with the intention of joining the ministry, and both fell under the influence of clergymen who happened to be leading scientists: in Clarke's case the Rev. Adam Sedgwick, Professor of Geology; and in Darwin's case, both Sedgwick and the Rev. John Stevens Henslow, Professor of Botany. Unlike Darwin, Clarke entered the church after leaving Cambridge, but he maintained a keen interest in geology for the rest of his life, most of which was spent in Australia. As we have seen, Clarke was much more enthusiastic about Darwin's theory than was his parishioner and churchwarden Conrad Martens. Interestingly, Clarke was also much more enthusiastic than Sedgwick, who wrote a scathing review of the *Origin*, and who set examination questions critical of his former student's theory.

A photograph of Charles Darwin taken towards the end of his life, c. 1880.

Apart from Clarke, Darwin's most ardent supporter in Australia was Gerard Krefft, who was curator of the Australian Museum from 1861 to 1874. Both Clarke and Krefft corresponded with Darwin, sending him various items of information, some of which were incorporated into later editions of the *Origin*.

In the years following publication of the *Origin*, Darwin continued his investigations into a wide variety of topics in natural history. The extent and breadth of his activity can be judged from the following ten books that he wrote during his remaining twenty-two years:[62]

On the Various Contrivances by Which British and Foreign Orchids are Fertilised by Insects, and on the Good Effects of Intercrossing (1862; 2nd edn, 1877)

On the Movements and Habits of Climbing Plants (Longman, Green, Longman, Roberts, & Green and Williams & Norgate, London, 1865; 2nd edn, John Murray, 1875)

The Variation of Animals and Plants under Domestication (2 vols, 1868; 2nd edn, 1875)

The Descent of Man, and Selection in Relation to Sex (2 vols, 1871; 2nd edn, 1874)

The Expression of the Emotions in Man and Animals (1872)

Insectivorous Plants (1875)

The Effects of Cross and Self Fertilisation in the Vegetable Kingdom (1876)

The Different Forms of Flowers on Plants of the Same Species (1877)

The Power of Movement in Plants (with Francis Darwin) (1880)

The Formation of Vegetable Mould, through the Action of Worms, with Observations on Their Habits (1881)

In addition to this remarkable output, Darwin also prepared four[63] revised editions of the *Origin* (1861, 1866, 1869, 1872).

Despite being increasingly troubled by ill health, as shown clearly in his letters to Syms Covington, Darwin lived to celebrate his 73rd birthday. Finally, on 19 April 1882, he died, and, as described in P. G. King's letter to Hallam Murray, he was buried in Westminster Abbey.

Summary of documents relating to Darwin's visit to Australia

....

A. Manuscript Sources

Detailed background information on the various types of manuscripts recording Darwin's observations during the *Beagle*'s voyage has been given in volume 1 of *Correspondence* (appendix II, pp. 545–8). Increasingly, these manuscripts are being made available via the website *The Complete Work of Charles Darwin Online*, at http://darwin-online.org.uk/. The following paragraphs list and briefly describe those parts of the whole manuscript record that relate directly to Darwin's observations in Australia. DAR numbers refer to material in the Darwin Archive within Cambridge University Library.

1. Field notebooks

As described in the Preface, these are the rough notes that Darwin made in the field. For the trip from Sydney to Bathurst, he started a new notebook known variously as notebook 1.3, notebook 14, or the Sydney–Mauritius notebook. For Hobart and King George Sound, there are no known notebook entries. Instead, there are similar types of notes written on various loose sheets of paper. In his book *Charles Darwin in Western Australia* (University of Western Australia Press, Nedlands, 1985), Dr P. Armstrong has argued convincingly that the loose sheet notes are the only field notes recorded by Darwin in these two places.

The Sydney–Mauritius notebook is now held in the Down House Museum, while the other field notes are held in the Darwin Archive at the Cambridge University Library. The Hobart notes are in DAR 40, fols. 45–9, the geological portions of which have been published in Banks, M. R. and Leaman, D., 'Darwin's field notes on the geology of Hobart town – a modern appraisal.' *Papers and Proceedings of the Royal Society of Tasmania* 133 (1999): 29–50. The King George Sound notes are scattered amongst zoological and geological notes in DAR 29, 31, and 38.

2. Zoological and geological diaries and notes

These are all housed in the Darwin Archive at the Cambridge University Library: Australian zoology in DAR 31, and Australian geology in DAR 38. More detailed descriptions are as follows.

DAR 31.2, pp. 346–8: Hobart lizards and snakes

DAR 31.2, p. 349: Conferva [phytoplankton] off the coast of Australia

DAR 31.2, pp. 363–6: Planaria at Hobart and afterwards

DAR 38.1, pp. 812–36: NSW geology, including itinerary (813)

DAR 38.1, pp. 837–57: Hobart geology (published in Banks, M. R., 'A Darwin manuscript on Hobart Town.' *Papers and Proceedings of the Royal Society of Tasmania* 105 [1971]: 5–19)

DAR 38.1, pp. 858–81: King George Sound geology

3. Catalogues of specimens

Lists of specimens together with some miscellaneous notes are contained in DAR 29.1 and DAR 29.3.

An extensive catalogue of insects collected in Australia is included in a manuscript entitled *Copy of Darwin's notes in reference to Insects collected by him* (in Syms Covington's hand), which is preserved in the Entomological Library of the Natural History Museum. For an informative published edition of this catalogue, see Smith, K. G. V., 'Darwin's insects: Charles Darwin's entomological notes.' *Bulletin of the British Museum (Natural History)* (Historical Series) 14 (1987): 1–143 and Smith, K. G. V., 'Supplementary notes on Darwin's insects.' *Archives of Natural History* 23(2) (1996): 279–86.

B. PUBLISHED SOURCES

The following list indicates the relatively small number of Darwin's Australian observations and collections that were published. As with the manuscripts, increasingly these publications are becoming available via the website *The Complete Work of Charles Darwin Online*, at http://darwin-online.org.uk/.

1. Books arising from the Beagle's voyage

Zoology, Part II, Living Mammalia (1839)

pp. 66–7 and plate 25: *Mus fuscipes*

pp. 67–8 and plate 34, figs 18a, b: *Mus gouldii*

Zoology, Part IV, Fish (1840–1842)

pp. 13–15: *Arripis georgianus*

pp. 18–20: *Helotes octolineatus*

pp. 33–5: *Platycephalus inops*

pp. 68–9 and plate 14: *Caranx declivis*

pp. 71–2: *Caranx georgianus*

pp. 82–3: *Dajaus diemensis*

pp. 137–8: *Platessa* sp.

pp. 156–7: *Aleuteres maculosus*

pp. 157–8: *Aleuteres velutinus*

p. 163: *Apistus* sp.

Zoology, Part V, Reptiles (1843)

p. 30 and plate 15, fig. 3: *Cyclodus casuarinae*

pp. 33–4 and plate 16, fig. 4: *Cystignathus georgianus*

Volcanic Islands (1844)

ch. VII: Geology of New South Wales, Van Diemen's Land, and King George's Sound

pp. 158–60: Palaeozoic shells from Van Diemen's Land, by G. B. Sowerby

pp. 161–9: Description of six species of corals, from the Palaeozoic formation of Van Diemen's Land, by W. Lonsdale

2. Scientific papers arising from the Beagle's voyage

Waterhouse, G. R. 'Descriptions of some of the insects brought to this country by C. Darwin.' *Transactions of the Entomological Society of London* 2(2) (1838): 131–5.

Waterhouse, G. R. 'Descriptions of some new species of exotic insects.' *Transactions of the Entomological Society of London* 2(3) (1839): 188–96, plate 17.

Babington, C. C. 'Dytiscidae Darwinianae; or, descriptions of the species of *Dytiscidae* collected by Charles Darwin, Esq., M.A., Sec. G.S. &c., in South America and Australia, during his voyage in H.M.S. *Beagle*.' *Transactions of the Entomological Society of London* 3(1) (1841): 1–17.

Saunders, W. W. 'Descriptions of new Australian *Chrysomelidae* allied to *Cryptocephalus*.' *Annals and Magazine of Natural History* 11, April 1843, 'Proc. Ent. Soc. Lond., for June 6, 1842', p. 317 only. Reprinted in *Journal of Proceedings of the Entomological Society of London*, Jan. 1840 to Dec. 1846, p. 70 only (in association with *Transactions of the Entomological Society of London* 4).

Lea, A. M. 'On some Australian Coleoptera collected by Charles Darwin during the voyage of the "*Beagle*".' *Transactions of the Entomological Society of London* 74(2) (1926): 279–88.

C. DARWIN'S AUSTRALIAN CORRESPONDENCE

In the absence of the telephone, Darwin was a very active writer of letters. By the time of publication of the *Calendar of the Correspondence of Charles Darwin, 1821–1882* (ed. F. Burkhardt and S. Smith, Garland, New York, 1985), almost 14 000 letters written either by or to Darwin were known. This number has since increased to around 14 700. The immense task of publishing each of these letters after detailed editing is being undertaken in the multi-volume *The Correspondence of Charles Darwin* (Cambridge University Press, Cambridge). Volume 1 of this monumental effort was published in 1985. Thirty or so volumes later, the project is expected to be completed in 2025. Four years after each letter is published in *Correspondence*, it is made freely available via the website of the Darwin Correspondence Project at http://www.darwinproject.ac.uk. For each of the letters not yet available, a summary is available on this website.

A total of 174 letters listed in the *Calendar* relate directly or indirectly to Australia. The first, in 1831, refers to consultations with Captain P. P. King prior to the *Beagle*'s departure. The last, in 1881, is a response from an Australian sheep farmer to an article on black sheep that Darwin had published in the scientific journal *Nature*.

The *Calendar* identification numbers of letters written by or to Darwin that relate directly or indirectly to Australia are:

123, 206, 218, 222, 289, 293, 294, 295, 298, 299, 301, 337, 362, 363, 394, 430, 468, 475, 513, 514, 515, 522, 685, 686, 700, 799, 803, 871, 901, 927, 940, 966, 970, 979, 1010, 1017, 1019, 1020, 1021, 1030, 1135, 1237, 1275, 1370, 1427, 1436, 1477, 1481, 1538, 1588, 1607, 1637, 1638, 1644, 1694, 1714, 1794, 1840, 1889, 1923, 2033, 2050, 2056, 2188, 2276, 2358, 2371, 2377, 2378, 2382, 2385, 2386, 2388, 2400, 2401, 2404, 2507, 2606, 2608, 2702, 2803, 2915, 2942, 2944, 2992, 3026, 3222, 3269, 3298, 3392, 3398, 3401, 3579, 3616, 3727, 3733, 3741, 3749, 3759, 3806, 3809, 4109, 4280, 5300, 5424, 5620, 5626, 5672, 5677, 5709, 5716, 5838, 5896, 5899, 5916, 5948, 6306, 6314, 6374, 6419, 6611, 6635, 6871, 7012, 8131, 8159, 8182, 8196, 8331, 8416, 8698, 8768, 8834, 8895, 8903, 8930, 8959, 8970, 8975, 9002, 9037, 9044, 9124, 9129, 9241, 9246, 9494, 9694, 10069, 10133, 10161, 10579, 10700, 10798, 10893, 10915, 10917, 10934, 11130, 11209, 11693, 11719, 12158, 12408, 12752, 12838, 12841, 12955, 13045, 13154, 13192, 13262, 13369, 13862.

Each of these letters can be located (in either complete or summary form) at http://www. darwinproject.ac.uk by searching for the *Calendar* number.

D. SYMS COVINGTON'S DIARY

Syms Covington recorded his own brief impressions of Australia at the time of Darwin's visit. These are now held in the Mitchell Library (ML), State Library of New South Wales, Sydney. The full reference is:

Covington, Syms – Diary of visit to Australia. p. 3 Linnean Society of NSW Records. 1826–1941, VII, Papers of Syms Covington, 1831–1836, 1839. ML MSS 2009/108 item 5.

An annotated transcript of Covington's diary is available at http://www.asap.uni melb.edu. au/bsparcs/covingto/.

E. REMINISCENCES OF P. G. KING (JUNIOR)

P. G. King (Junior) recorded his reminiscences on three occasions. The manuscripts or copies of these are now held in the Mitchell Library (ML), State Library of New South Wales. The full references are:

King, P. G., the younger – Reminiscences re Charles Darwin 1831–1836. Sent to W. A. H. Hallam Murray October 17, 1892. ML FM4/6900, frames 338–46.

King, P. G., the younger – Autobiography 1894. ML FM4/6900, frame 320.

P. G. King (the younger)'s autobiography. In Memoirs of Governor King, Admiral King and an autobiography by Philip Gidley King, the younger. ML C770.

ILLUSTRATION AND MAP ACKNOWLEDGEMENTS

....

The authors are very grateful to all the institutions and individuals acknowledged below, for permission to reproduce the material as detailed below. Illustrations without acknowledgement were obtained from material not subject to copyright, or were provided by the authors. Every effort has been made to obtain permission where it is required. The authors would be very grateful for advice on any omissions or errors, which they will be pleased to correct at the first opportunity.

Front cover: Charles Darwin in 1840. Watercolour portrait by George Richmond. © The Gallery Collection/Corbis. *H.M.S. Beagle, Sydney*. Pencil sketch by Conrad Martens. Dated 'May 9th.39'. In his *Sketches in Australia*. 25.2 cm × 20.3 cm. Reference: ZPXC295, fo. 40. Mitchell Library, State Library of New South Wales.

Endpapers: The round-the-world route of the *Beagle*. Loose map from the *Narrative*, appendix to vol. 2.

Page vi: Conrad Martens, [Sydney Harbour], nla.pic-an2390645, National Library of Australia.

Page x: *Jamieson Valley, NSW, Looking Towards King's Tablelands*. Undated watercolour by Conrad Martens. Courtesy of Mr R. P. Harbig. Based on a sketch entitled *View near the Weatherboard, Bathurst Road*, 23 May [1838], now in the Dixson Library (DL PX24, fo. 37).

Page 3: Capt. Robert FitzRoy. Drawn by Philip Gidley King (Junior). Dated 1838, but most likely drawn in 1835 or early 1836. 14 cm × 19.5 cm unmounted. Reference: ZC767,

p. 68. Mitchell Library, State Library of New South Wales.

Page 5: Francis Beaufort. Crayon sketch by William Brockendon, 13.3.1838. 35.6 cm × 26.0 cm. Reference: 2515/90. National Portrait Gallery, London.

Page 7: 'Figures to denote the force of the wind', 'Letters to denote the state of the weather', and meteorological data collected by the *Beagle* during her stay in Sydney. *Narrative*, appendix to vol. 2, pp. 40, 41, 51 (12 Jan. to 30 Jan.).

Page 10 (left): Augustus Earle, Solitude, watching the horizon at sun set, in the hopes of seeing a vessel, Tristan de Acunha [i.e. da Cunha] in the South Atlantic, nla.pic-an2818137, National Library of Australia. 17.5 cm × 25.7 cm.

Page 10 (right): Sketch of Philip Gidley King (Junior), as a boy of eight, on board the *Beagle*. Approx. 12.5 cm × 14 cm. Reference: PXC767, fo. 142. Mitchell Library, State Library of New South Wales.

Page 12: The *Beagle*'s quarter-deck and poop cabin. Drawn from memory by Philip

Gidley King (Junior) in 1890. 33 cm × 22 cm. Reference: A1977. King papers, vol. 2, p. 813. Mitchell Library, State Library of New South Wales.

Page 13: A diagrammatic section of the *Beagle*. Drawn from memory by Philip Gidley King (Junior) in 1890. 33 cm × 22 cm. Reference: A1977. King papers, vol. 2, p. 811. Mitchell Library, State Library of New South Wales.

Page 14: Self-portrait by Conrad Martens. Signed and dated 'C.M. June 1834', with note 'On board HMS *Beagle*'. 10 cm × 13.5 cm. Reference: Pd279. Dixson Library, State Library of New South Wales.

Page 17: *Beagle off Mt Sarmiento*. Oil by Conrad Martens. Current owner/location unknown.

Page 19: Chronometer readings, New Zealand to Sydney. In *Narrative*, appendix to vol. 2, p. 338.

Page 22 (top): *Entrance to Port Jackson N. S. Wales*. Pencil sketch by Conrad Martens. Dated 'April 17./35'. In his *Album of Pencil Sketches in NSW*. 28.5 cm × 19 cm. Reference: ZDLPX24, fo. 1. Dixson Library, State Library of New South Wales.

Page 22 (bottom): *Sydney Lighthouse*. Pencil sketch by Conrad Martens. Dated 'Dec 6./35'. In his *Album of Pencil Sketches in NSW*. 30 cm × 20 cm. Reference: ZDLPX24, fo. 21. Dixson Library, State Library of New South Wales.

Page 23: *Sydney* 1836. Watercolour by Conrad Martens. In his *Views Mainly of NSW*, vol. 5. 62.5 cm × 44 cm (mounted, 63 cm × 44.5 cm). Reference: DDGD11, fo. 12. Dixson Galleries, State Library of New South Wales.

Page 24: *Harbour Scene Showing Fort Macquarie*. Watercolour by Conrad Martens. Dated 1836. In his *Views Mainly of NSW*. 48 cm × 32.5 cm (mounted, 58.5 cm × 44.1 cm). Reference: ZDGD8, fo. 5. Dixson Galleries, State Library of New South Wales.

Page 25 (top): Shipping intelligence. *Sydney Gazette*, vol. 34, no. 2759, Thurs. 14 Jan., 1836, p. 2.

Page 25 (bottom): *H.M.S. Beagle, Sydney*. Pencil sketch by Conrad Martens. Dated 'May 9th.39'. In his *Sketches in Australia*. 25.2 cm × 20.3 cm. Reference: ZPXC295, fo. 40. Mitchell Library, State Library of New South Wales.

Page 27 (top): *Bridge Street*. Watercolour by Conrad Martens. Signed and dated 'C. Martens Sydney 1839' lower right. Title on mount below. 44.8 cm × 64.8 cm. Reference: ZDG V*/SP COLL/MARTENS 7. Dixson Galleries, State Library of New South Wales.

Page 27 (bottom): 'Map of the town of Sydney, 1836. Drawn at the Surveyor General's Office. Lithographed for J. Mudie's Work on N. S. Wales by J. and C. Walker'. Reference: M2 811.16/1836/2. Mitchell Library, State Library of New South Wales.

Page 29: *In the Domain*. Pencil sketch by Conrad Martens. Dated 'Sep 25.1835'. In his *Sydney -&c* sketchbook. 21.5 cm × 12.5 cm. Reference: ZPXC391, fo. 16. Mitchell Library, State Library of New South Wales.

Page 30: Title section from 'Map of the Colony of New South Wales' compiled by T. L. Mitchell, Surveyor General. Sydney; drawn by T. L. Mitchell; engraved by John Carmichael; republished London, 1834. Reference: MT4 811/1834/1. Mitchell Library, State Library of New South Wales.

Page 31: Front cover and first page of field notebook carried by Darwin during his trip to Bathurst. Actual size: front cover 9.3 cm × 7.8 cm; page 8.8 cm × 6.7 cm. Courtesy of Mr P. Titheradge, former curator, Down House Museum.

Page 33–4: Itinerary; comprising roads throughout New South Wales. From the *New South Wales Calendar and General Post Office Directory, 1835* (Stephens and Stokes, Sydney 1835), pp. 130–4 (from

Sydney [zero miles] to Parramatta [15 miles]). Reference: 991.01/N. Mitchell Library, State Library of New South Wales.

Page 34 (bottom): *View of Parramatta.* Watercolour by Conrad Martens. Dated 1838. 66 cm × 45 cm (mounted 76 cm × 52 cm). Reference: ZDLPg15. Dixson Library, State Library of New South Wales.

Page 36: *A Government Jail Gang, Sydney, N. S. Wales.* Lithograph by Augustus Earle. In his *Views in New South Wales and Van Diemen's Land* (J. Cross, Holborn, London, 1830), 30 cm × 27.5 cm. Reference: F981E, copy no 1. Mitchell Library, State Library of New South Wales.

Page 37: Great Western Road, from Sydney to Emu Ford. *New South Wales Calendar and General Post Office Directory, 1835,* pp. 146–9 (from 14 ½ miles to 34 ¾ miles). Reference: 991.01/N. Mitchell Library, State Library of New South Wales.

Page 38: *Penrith Road.* Dated 'Dec 26/35'. Pencil sketch by Conrad Martens in his *Sydney -&c* sketchbook. 21.5 cm × 12.5 cm. Reference: ZPXC391, fo. 20. Mitchell Library, State Library of New South Wales.

Page 40: *Bluegum.* Watercolour by Conrad Martens. 17.5 cm × 25.5 cm (mounted, 36.5 cm × 45.7 cm). Reference: SV*SP COLL/3. Mitchell Library, State Library of New South Wales.

Page 45 (top): *Emu Ferry, Great Western Road.* Pencil sketch by Conrad Martens. Dated 'May.15.35'. 29 cm × 19.5 cm. In his *Album of Sketches in NSW.* Reference: ZDLPX24, fo. 3. Dixson Library, State Library of New South Wales.

Page 45 (bottom): Modern (c. 1986) photograph of Emu Ferry.

Page 46: 'Survey of Part of Emu Plains showing the new and old line of Road to the Pilgrim', 16 July 1832. In Major Mitchell's *Report upon the Progress Made in Roads and in the Construction of Public Works from the Year 1827 to June 1855,* illustration facing p. 33. Reference: A331. Mitchell Library, State Library of New South Wales.

Page 47: Conrad Martens, [Lennox Bridge, Lapstone Hill, Mitchell Pass, near Penrith, N.S.W. 1835]. 46.1 cm × 66.9 cm, nla.pic-an2390736, National Library of Australia.

Page 48: Great Western Road. *New South Wales Calendar and General Post Office Directory, 1835,* pp. 184–185 (from 34 ½ miles to 69 miles). Reference: 991.01/N. Mitchell Library, State Library of New South Wales.

Page 49: Modern (c. 1986) photograph of the oak tree on the site of the Weatherboard Inn, planted in 1936 to commemorate the centenary of Darwin's visit.

Page 50: Modern (c. 1986) photograph of Jamison Creek.

Page 51 (top): *Jamieson Valley, NSW, Looking Towards King's Tablelands.* Undated watercolour by Conrad Martens. Courtesy of Mr R. P. Harbig. Based on a sketch entitled *View near the Weatherboard, Bathurst Road,* 23 May [1838], now in the Dixson Library (DL PX24, fo. 37).

Page 51 (bottom): Modern (c. 1986) photograph of the Jamison Valley.

Page 53: Gardner's Inn. Photograph in the possession of Tom Bennet, 1963. Reference: SPF/Blackheath – Hotels – Gardiners Inn. Mitchell Library, State Library of New South Wales.

Page 54: Modern (c. 1986) photograph of Gardner's Inn. Courtesy of Lowan Turton.

Page 56: 'Wide views of the Grose Valley from Govetts Leap, Blue Mountains', modern (Jan 2006) photograph. Courtesy of John Turner.

Page 58: *Valley of the Grose.* Oil by Conrad Martens. c. 1839. 89 cm × 62.2 cm. Current owner/location unknown.

Page 61: *Victoria Pass.* Pencil sketch by Conrad Martens. Dated 'May 24-38'. 25.2 cm × 18 cm. Reference: ZDLPX24, fo. 42. Dixson Library, State Library of New South Wales.

Page 62: Great Western Road (continued). *New South Wales Calendar and General Post Office Directory, 1835*, pp. 186–7 (from 71 miles to 80 miles). Reference: 991.01/N. Mitchell Library, State Library of New South Wales.

Page 63: Road to Mudgee and Pandora's Pass. *New South Wales Calendar and General Post Office Directory, 1835*, pp. 194 (from 80 ¼ miles to 88 miles). Reference: 991.01/N. Mitchell Library, State Library of New South Wales.

Page 64: Modern (c. 1986) photograph of the site of the Wallerawang homestead.

Page 66: New South Wales Rat Kangaroo. From Gould, J., *The Mammals of Australia* (published by the author, London, 1845–1863), vol. II, plate 67. Courtesy of Hank Ebes.

Page 69: Crimson Rosella, Eastern Rosella, and Australian King Parrot. From Gould, J., *The Birds of Australia* (published by the author, London, 1840–1848), vol. v, plates 23, 27, and 17, respectively. Courtesy of Hank Ebes.

Page 70 (top left): Sulphur-crested Cockatoo. From Gould, J., *The Birds of Australia* (published by the author, London, 1840–1848), vol. v, plate 1. Courtesy of Hank Ebes.

Page 70 (top right): Australian Raven. From Gould, J., *The Birds of Australia* (published by the author, London, 1840–1848), vol. iv, plate 18. Courtesy of Hank Ebes.

Page 70 (bottom): Australian Magpie and Pied Currawong. From Gould, J., *The Birds of Australia* (published by the author, London, 1840–1848), vol. ii, plates 46 and 42, respectively. Courtesy of Hank Ebes.

Page 72: *Cox's River*. Watercolour by Conrad Martens. Late 1830s. Current owner/location unknown.

Page 73: Platypus. From Gould, J., *The Mammals of Australia* (published by the author, London, 1845–1863), vol. I, plate 1. Courtesy of Hank Ebes.

Page 75 (left): Photograph of Ant-lion pits. Courtesy of Densey Clyne.

Page 75 (right): Typical larva of Myrmeleontidae, the dominant family of lacewings. Photograph courtesy of Dr G. Holloway.

Page 80–1: Great Western Road (continued). *New South Wales Calendar and General Post Office Directory, 1835*, pp. 187–90 (from 80 ¾ miles to 114 miles). Reference: 991.01/N. Mitchell Library, State Library of New South Wales.

Page 82: Plaque in Machattie Park at Bathurst, commemorating Darwin's visit. Courtesy of Graham Lupp.

Page 83 (top): *Bathurst from the West*. Pencil sketch by Conrad Martens. 30.5 cm × 18.5 cm (mounted, 34.5 cm × 46.5 cm). Reference: ZSV*SP COLL MARTENS/1. Mitchell Library, State Library of New South Wales.

Page 83 (bottom): Plan of the Town of Bathurst, TL Mitchell. [19 Jan.] 1833 from State Records NSW: Lands Department; NRS 8224. Maps and plans arranged by the Mitchell Library's map classification system, 1760–1920, [Map 117]. Facsimile of the original held by State Records NSW. Courtesy of State Records NSW and Department of Lands NSW.

Page 85 (top): Holy Trinity Church, Kelso, as it appeared in the 1890s. Photograph from Burton, B., Fields, E., Neumann, D. and White, S., *Holy Trinity Church, Kelso, 1835: An Historical Analysis* (University of NSW, Studies of Historical Buildings, unpublished report). Reference: PXD267, fig. 8. Mitchell Library, State Library of New South Wales.

Page 85 (bottom): Modern photograph of Holy Trinity Church, Kelso. Photograph from Burton, B., Fields, E., Neumann, D. and White, S., *Holy Trinity Church, Kelso, 1835: An Historical Analysis* (University of NSW, Studies of Historical Buildings, unpublished report). Reference: PXD267, fig. 2. Mitchell Library, State Library of New South Wales.

Page 91 (top): *View of Dunheved, New South Wales*. Watercolour by Conrad Martens. 31.5

cm × 47 cm (sight) in original nineteenth century wood veneer frame (probably English), 46 cm × 61.7 cm (outside meas.), with gilded inner slip. Reference: ML 1140. Mitchell Library, State Library of New South Wales.

Page 91 (bottom): 'Where C & P Leslie lived when first married Dunheved South Creek rented fr Admiral King'. Pencil sketch probably by Philip Gidley King (Junior), 1840. 29.5 cm × 18.5 cm (mounted, 38 cm × 55.5 cm). Reference: PX*D379, vol. 2, fo. 26. Mitchell Library, State Library of New South Wales.

Page 92: *Admiral Philip Parker King*. Oil painting. Artist unknown. 49.5 cm × 59.3 cm (framed, 67.5 cm × 77.3 cm). Reference: ML 11. Mitchell Library, State Library of New South Wales.

Page 93: *Beagle, an Australian Bred Horse, by Skeleton, the Property of Captn P.P. King, R.N.* Watercolour by James Lethbridge Templar. Signed and dated 'JLT 1839 Decr.' 27 cm × 34 cm (mounted, 35.5 cm × 46 cm). Reference: SV*/HORS/5. Mitchell Library, State Library of New South Wales.

Page 94: Portrait of Mr H. H. Macarthur, 1800–1820? 30.5 cm × 26 cm (framed, 35.7 cm × 31.3 cm). Reference: ZML145. Mitchell Library, State Library of New South Wales.

Page 95: *The Vineyard Estate, Subiaco. 12.8.39. Near Parramatta*. Oil by Conrad Martens. 61 cm × 43.5 cm (framed, 73.6 cm × 54 cm). Reference: ZML48 Pict. Stor. Mitchell Library, State Library of New South Wales.

Page 96: Front View of The Vineyard (Subiaco) in 1961, just prior to its demolition. Reference: UNSW central file 12170. Courtesy University Archives, University of New South Wales.

Page 97 (left): Modern (c. 1986) photograph of the colonnade from Vineyard, as re-erected beside the Village Green Oval, University of New South Wales.

Page 97 (right): Main Entrance to Vineyard (Subiaco), taken just prior to its demolition

in 1961. Reference: UNSW central file 12170. Courtesy University Archives, University of New South Wales.

Page 100: Plan and side elevation of observatory at Parramatta. In Richardson, W., *A Catalogue of 7385 stars, chiefly in the Southern Hemisphere, Prepared from Observations made in the Years 1822, 1823, 1824, 1825 and 1826, at the Observatory at Parramatta, New South Wales, founded by Lieutenant-General Sir Thomas Makdougall Brisbane, K.C.B. F.R.S., President of the Royal Society of Edinburgh. The Computations made, and the catalogue constructed, by Mr William Richardson, of the Royal Observatory at Greenwich* (His Majesty's Stationery Office, London, 1835), illustration facing title page.

Page 101: Modern (c. 1986) photograph of the remains of the observatory at Parramatta.

Page 102: Modern photographs of Old Government House, Parramatta, and its dining room (13 March 2006). Courtesy of the National Trust of Australia (NSW) and Christopher Shain, respectively.

Page 109: Second page of Conrad Martens's account book, recording the sale of two paintings to Charles Darwin, on 17 and 21 January, 1836. Reference: CYMS142. Dixson Library, State Library of New South Wales.

Page 110: *Hauling the Boats up the Rio Santa Cruz*. Watercolour by Conrad Martens. Courtesy of Mrs R. G. Barnet.

Page 111: *The Beagle in Murray Narrow, Beagle Channel*. Watercolour by Conrad Martens. Courtesy of Mr George Darwin.

Page 112: Medallion modelled by Josiah Wedgwood from Sydney Cove clay sent to England by Captain Arthur Phillip. Etruria, Staffordshire. Reference: ML P*68. Mitchell Library, State Library of New South Wales.

Page 117: *Leptosomus acuminatus* (order Coleoptera, family Curculionidae [weevils]), one of the beetles collected by Darwin in Sydney. From Waterhouse, G.R., Descriptions of some new species of exotic insects. *Transactions of the Entomological Society of London* 2(3) (1839): 188–96, plate 17, as reprinted in Smith, K. G. V., 'Darwin's insects: Charles Darwin's entomological notes,' *Bulletin of the British Museum (Natural History) Historical Series* 14(1) (1987): 1–143, p. 93. Courtesy of K.G.V. Smith and the Natural History Museum.

Page 122: 'Entrance to the River Derwent, Tasmania, showing the lighthouse'. Pencil sketch by Syms Covington. 22.5 cm × 17 cm (mounted, 38 cm × 60 cm). Reference: PXD41, fo. 2b. Mitchell Library, State Library of New South Wales.

Page 124: Hobart warehouses and bondstores, Salamanca Place. Photograph; gelatin silver. Reference: H98.252/2230; jc020548. J. T. Collins Collection, La Trobe Picture Collection, State Library of Victoria.

Page 125: Panorama of Hobart Town. Frontispiece plus title page to booklet entitled *Description of a View of Hobart Town, Van Diemen's Land, and showing the Surrounding Country, Now Exhibiting at the Panorama, Strand. Painted by the Proprietor, Mr R. Burford* (London 1831). Frontis. 43.5 cm × 26.5 cm. Reference: 986.1/2A1. Mitchell Library, State Library of New South Wales.

Page 126–7: Panorama of Hobart Town, c. 1826. Six watercolours by Augustus Earle. Each 54 cm × 36.5 cm (mounted, 70.5 cm × 50 cm). Reference: ZDGD14, fos. 1–6. Dixson Galleries, State Library of New South Wales.

Page 134: Photograph of *Polyporella internata* (= *Fenestella internata*). Fossil bryozoan from Bundella Mudstone, Lower Permian, shoreline, Porter Hill, Hobart. Courtesy of Dr M. R. Banks.

Page 137: Title and portion of 'Map of Van Diemen's Land' by George Frankland, London, engraved and published by J. Cross, 18 Holborn, London. March 1st, 1839. Reference: ZM4/880/1839/1. Mitchell Library, State Library of New South Wales.

Page 139: *Secheron*. 12 July 1846. Pencil sketch by Thomas Chapman. Tasmanian Museum and Art Gallery.

Page 141: Modern photograph of Stephenville, St Michael's Collegiate School, Macquarie Street, Hobart. Courtesy of Louise Lindsay and St Michael's Collegiate School.

Page 142: 'Alfred Stephen, April 1839, Aet 37, Hobart Town, Van Diemen's Land'. Pencil on paper, attributed to T. G. Wainewright. Courtesy of Allport Library and Museum of Fine Arts, State Library of Tasmania.

Page 150–1: Itinerary of the route from Hobart to New Norfolk. In *Van Diemen's Land Almanack*, 1833, pp. 138–40. Reference: 996.01/V. Mitchell Library, State Library of New South Wales.

Page 153: Oak skink, *Cyclodus casuarinae* (now called *Tiliqua casuarinae*). From Darwin, C.R. (ed.), *The Zoology of the Voyage of H.M.S. Beagle* (Smith, Elder and Co., London, 1843), Part V: *Reptiles*, by Thomas Bell; plate 15, fig. 3.

Page 155: *Eucharis iello*, a gall-forming fig wasp (order Hymenoptera, superfamily Chalcidoidea), one of the insects collected by Darwin during the *Beagle*'s stay in Hobart. From Walker, F., *Entomologist* 1 (1840–2): plate P, as reprinted in Smith, K.G.V., 'Darwin's insects: Charles Darwin's entomological notes,' *Bulletin of the British Museum (Natural History) Historical Series* 14 (1) (1987): 1–143, p. 92. Courtesy of K. G. V. Smith and the Natural History Museum.

Page 158–9: *Panoramic View of King George's Sound, Part of the Colony of Swan River*. Coloured aquatint, drawn by Lieut. R. Dale, engraved by R. Havell. London, 1834. Reference: PXB3. Mitchell Library, State Library of New South Wales.

Page 159 (top): 'Plan of King George's Sound and its Harbours, shewing the Boundaries and Limits of the Townsite of Albany and Crown Reserves, 1835'. Reproduced by permission of Western Australian Land

Information Authority, C/L Albany 30M; www.landgate.wa.gov.au.

Page 160: Grass-tree, *Kingia australis* (a20342). Courtesy of the Australian Plant Image Index, Australian National Botanic Gardens; © M. Fagg, Australian National Botanic Gardens.

Page 162 (top): 'King George's Sound'. Pencil sketch by Syms Covington. 56 cm × 19 cm. Reference: PXD41, fo. 2c. Mitchell Library, State Library of New South Wales.

Page 162 (bottom): Article commencing 'On Wednesday 9th March . . .', *Perth Gazette*, vol. 4, no. 169, Sat. 26 March 1836, p. 674c.

Page 163: Modern (February 2008) photograph of Old Farm Strawberry Hill. Collection: National Trust of Australia (WA).

Page 166: *Corroboree, Southwest of Western Australia*, c. 1843. Watercolour by Richard Atherton ffarrington. Reference: 435828 18 3M A1. State Art Collection, Art Gallery of Western Australia.

Page 169 (top): Australian bush rat, *Mus fuscipes* (now called *Rattus fuscipes*). From Darwin, C.R. (ed.), *The Zoology of the Voyage of H.M.S. Beagle* (Smith, Elder and Co., London, 1843), Part II: *Mammalia*, by George R. Waterhouse; plate 25.

Page 169 (bottom): Southern frog, *Cystignathus georgianus* (now called *Crinia georgiana*). From Darwin, C.R. (ed.), *The Zoology of the Voyage of H.M.S. Beagle* (Smith, Elder and Co., London, 1843), Part V: *Reptiles*, by Thomas Bell: plate 16, fig. 4.

Page 171 (top): Jack mackerel, *Caranx declivis* (now called *Trachurus declivis*). From Darwin, C.R. (ed.), *The Zoology of the Voyage of H.M.S. Beagle* (Smith, Elder and Co., London, 1843), Part IV: *Fish*, by Leonard Jenyns: plate 14.

Page 171 (bottom): *Eucharis volusus*, a gall-forming fig wasp (order Hymenoptera, superfamily Chalcidoidea). Collected by Darwin in the vicinity of King George Sound. From Walker, F., *Entomologist* 1 (1840–2): plate P, as reprinted in Smith, K.

G. V., 'Darwin's insects: Charles Darwin's entomological notes,' *Bulletin of the British Museum (Natural History) Historical Series* 14 (1) (1987): 1–143, p. 92. *Allelidea ctenostonoides*, a flower-dwelling beetle (order Coleoptera, family Melyridae) and *Alleloplasis darwinii*, a plant-hopper (order Hemiptera, family Issidae). Collected by Darwin in the vicinity of King George Sound. From Waterhouse, G.R., Descriptions of some new species of exotic insects, *Transactions of the entomological society of London* 2 (3) (1839): 188–96, plate 17, as reprinted in Smith, K. G. V., 'Darwin's insects: Charles Darwin's entomological notes,' *Bulletin of the British Museum (Natural History) Historical Series* 14 (1) (1987): 1–143, p. 93. Courtesy of K. G. V. Smith and the Natural History Museum.

Page 174: Darwin's last comments on Australia, as recorded in his diary. From *The Journal of a Voyage in HMS Beagle* (Genesis Publications in association with ANZ Book Co. Pty Ltd, London and Sydney, 1979).

Page 176: Photograph of Syms Covington. Courtesy of Joy Sirl.

Page 184: Modern (c. 1986) photograph of Syms Covington's house.

Page 188 (top): View of Conrad Martens's House, Rockleigh Cottage, at St Leonards, 1851. Pencil drawing by Conrad Martens. 12.4 cm × 20.6 cm. Unsigned. Dated 'Feb 28th 1851'. Untitled. Reference: PX33, fo. 20. Dixson Library, State Library of New South Wales.

Page 188 (bottom): Old St Thomas's Church, North Sydney, 1845? Watercolour by Conrad Martens. 17.7 cm × 25 cm. Unsigned. Undated. Untitled. Reference: SV*/SP COLL/MARTENS 15. Mitchell Library, State Library of New South Wales.

Page 190: *View of Brisbane (in 1851)* 1862. Watercolour and gouache over pencil on wove paper, by Conrad Martens. 31.8 cm × 51.3 cm. Gift of Leonard Darwin 1913. Queensland Art Gallery.

Page 193: *View from Bulimba*. Pencil sketch by Conrad Martens. 18.9 cm × 30.2 cm. Unsigned. Undated. Reference: PXC301, fo. 7. Mitchell Library, State Library of New South Wales.

Page 194: *Tahlee*. Pencil sketch by Conrad Martens. Dated 'April 23./41-', and titled at lower left. Folio 59 in his *Pencil Sketches of New South Wales*. 18.2 cm × 28.1 cm. Unsigned. Reference: ZDLPX24. Dixson Library, State Library of New South Wales.

Page 196: *The funeral of Rear Admiral Phillip Parker King, 1856*. Watercolour. Signed 'C. Martens. Sydney. 1856'. 43.5 cm × 64 cm; inside frame 69.5 cm × 89.5 cm. Reference: ML 994. Mitchell Library, State Library of New South Wales.

Page 197: [Stroud House, NSW, ca 1854]. Watercolour by Conrad Martens. 29.6 cm × 43.2 cm. Reference: ZV*/SP COLL/ MARTENS 38. Mitchell Library, State Library of New South Wales.

Page 198: *Goonoo Goonoo*, 1861. Pencil sketch by G. B. G. King. 28 cm × 18.7 cm inside frame lines. Signed, dated and titled 'G.B.G. King Goonoo Goonoo Sept. 10th. 1861. N.S.W.' at lower right. Reference: SV1B/25. Dixson Galleries, State Library of New South Wales.

Page 205: Modern photographs of the gravestones of Philip Gidley King (Junior) and his wife Elizabeth in the churchyard of St Mary Magdalene, Great Western Highway, St. Marys.

Page 209: Photograph of Charles Darwin, aged approx. 71, taken by Elliott and Fry, on the verandah at Down House, c. 1880. Reproduced by kind permission of Syndics of Cambridge University Library.

....

The maps listed below have all been drawn by Tony Fankhauser.

Page 34: Route from Sydney to Parramatta
Page 37: Great Western Road from Parramatta to Emu Ferry
Page 48: Great Western Road from Emu Ferry to Blackheath
Page 49: Wentworth Falls area of the Blue Mountains
Page 62: Great Western Road from Blackheath to Hassan's Walls
Page 63: Darwin's detour to Wallerawang
Page 81: Great Western Road from Wallerawang to Bathurst
Page 88: Route of Darwin's return trip from Bathurst to Old Bowenfels
Page 93: Route of Darwin's return trip from Dunheved to Sydney
Page 129: Tasmania, showing places mentioned in the text
Page 132: Hobart and environs, showing places visited by Darwin
Page 161: Albany and environs, showing places visited by Darwin

NOTES

....

PREFACE

1 Owned by English Heritage. Listed as EH1.3 in the Darwin Manuscript Catalogue [http://darwin-online.org.uk/content/record?itemID=EH1.3]; transcription available at http://darwin-online.org.uk/content/frameset?viewtype=text&itemID=EH1.3&pageseq=1.

2 Complete transcriptions of most of these sources are now available from *The Complete Works of Charles Darwin Online* http://darwin-online.org.uk/

3 Owned by English Heritage. Listed as EHBeagleDiary in the Darwin Manuscript Catalogue [http://darwin-online.org.uk/content/record?itemID=EHBeagleDiary]; transcription by Kees Rookmaaker available at http://darwin-online.org.uk/content/frameset?viewtype=text&itemID=EHBeagleDiary&pageseq=1. The Diary was published in edited form by Barlow, 1933, and completely (with a new transcription) by Darwin's great grandson Richard Darwin Keynes, 1988. The Keynes' transcription is available at http://darwin-online.org.uk/content/frameset?itemID=F1925&viewtype=text&pageseq=1

4 *Correspondence* vol. 2, p. 198.

5 *Correspondence* vol. 2, p. 198. On page 34 of Freeman, 1977, it is explained that the only difference between volume III of the *Narrative* and the 'reissued' separate book is that in the latter, some of the preliminary pages have been discarded. The actual text is exactly the same in both versions. The separate book was so popular that it was reprinted the following year. Several facsimiles of this edition have been published, e.g. in 1952 by Hafner Publishing Company, New York. The volume has been republished in Barrett and Freeman, 1986, vols 2 and 3.

6 http://www.asap.unimelb.edu.au/bsparcs/covingto/contents.htm

ACKNOWLEDGEMENTS

1 We appreciate that some of the people in this list are no longer with us, but it is important to acknowledge their contribution.

LIST OF KEY REFERENCES

1 See also Preface note 3. A facsimile was published in 1979 by Genesis Publications, Guildford, Surrey, in association with the ANZ Book Company, Brookvale, New South Wales, under the title of *The Journal of a Voyage in HMS Beagle*. A transcription of the Diary by Darwin's grandaughter (Barlow, 1933) was republished in Barrett and Freeman, 1986, vol. 1. While Barlow's transcription is generally excellent, her interpretation of certain difficult words or passages in the Australian section differs from ours. Consequently, the Diary extracts in the present book are taken from our own transcription of the facsimile.

2 The text and/or page images of all three volumes are available at http://darwin-online.org.uk.

3 The 1845 *Journal* has been reprinted and republished on many occasions. Although some of the John Murray reprints were described as new editions and had different titles, for all practical purposes the text has always remained the same as in the original 1845 edition. Even when Darwin added a one-page postscript (really a brief list of errata) to the 1860 reprint, and the book was retitled *A Naturalist's Voyage*, the main text was not altered. For further details on all reprints and republications, including translation, up to 1972, see Freeman, 1977. The text and/or page images of the 1845, 1860 and 1890 editions are available at http://darwin-online.org.uk.

4 This volume has been republished on many occasions. For a complete list up to 1972, see Freeman, 1977, pp. 58–63. Since then, it has been republished in Barrett and Freeman, 1986, vol. 7. The text and/or page images are available at http://darwin-online.org.uk.

5 This volume has been republished on many occasions. For a complete list up to 1972, see Freeman, 1977, pp. 58–63. Since then, it has been republished in Barrett and Freeman, 1986, vol. 8. The text and/or page images are available at http://darwin-online.org.uk.

6 This volume has been republished on many occasions. For a complete list up to 1972, see Freeman, 1977, pp. 58–63. Since then, it has been republished in Barrett and Freeman, 1986, vol. 9. Page images are available at http://darwin-online.org.uk.

7 In contrast with the three volumes on geology, the *Zoology* was neither reprinted nor republished for many years. In 1980, a handsome limited-edition facsimile was published by Nova Pacifica, Wellington, New Zealand. The entire work became available with the publication of Barrett and Freeman, 1986, vols 4 (Fossil Mammalia and Living Mammalia), 5 (Birds) and 6 (Fish and Reptiles). The text and/or page images of all volumes are available at http://darwin-online.org.uk.

1 INTRODUCTION

1 The first given name of P. G. King (Senior) and his grandson P. G. King (Junior) was 'Philip', whereas the first given name of P. P. King was 'Phillip'.

2 Pike, 1967, pp. 61–4.

3 The official account of the voyage was given in *Narrative*, vol. I. The parts of the story that concern FitzRoy are retold in Mellersh, 1968, chs. 2 and 3; and in Gribbin and Gribbin, 2003, ch. 5.

4 For FitzRoy's own version of this story, see *Narrative*, vol. II and also *Narrative*, vol. I. For modern interpretations, see Mellersh, 1968, chs. 4 and 5; and Gribbin and Gribbin, 2003, ch. 4. See also Hazlewood, 2000.

5 *Narrative*, vol. II; Mellersh, 1968, chs 4 and 5; Gribbin and Gribbin, 2003, ch. 4.

6 Beaufort's instructions are reprinted in *Narrative*, vol. II, pp. 24–41.

7 The account given here is derived from Bruton, 1968, pp. 82–93; Carrington, 1939, pp. 189–90; and Raper, 1914. See also Sobel, 1995.

8 A complete list of the chronometers on board each vessel is given in *Narrative*, appendix to vol. II, pp. 318–19.

9 Full details of the chronometers on board are given in *Narrative*, appendix to vol. II, p. 325.

10 *Narrative*, appendix to vol. II, p. 331. It should be noted that FitzRoy's observations were not without their own errors, as FitzRoy himself was quick to point out (p. 345): 'It ought to be clearly stated, however, that the sum of all the parts which form the chain amounts to more than twenty-four hours, therefore error must exist somewhere.' (His total was 24 hours and 33 seconds; p. 345.) A full discussion of the chronometrical results for the first and second *Beagle* voyages is given in *Narrative*, appendix to vol. II, item

55, pp. 318–52. For a critical evaluation of FitzRoy's longitude determinations, see Auwers, 1884.

11 *Narrative*, vol. II, p. 37.

12 A full account of Beaufort's life and achievements is given in Friendly, 1977.

13 Various opinions have been expressed about the exact reason for Darwin being invited to travel on the *Beagle*. The present account is based largely on Gruber, 1969, and Burstyn, 1975; and on the original letters in *Correspondence*, vol. 1.

14 *Narrative*, vol. II, pp. 18–19.

15 Information on Darwin's Edinburgh days is based largely on material in *Correspondence*, vol. 1; and in Barlow, 1958. A transcription of manuscript notes describing the discoveries that he presented to the Plinian Society is included in Barrett, 1977, pp. 285–91.

16 *Correspondence*, vol. 1; Barlow, 1958.

17 *Narrative*, vol. II., p. 19.

18 When he was baptized, Earle's surname was spelt like that of his father, i.e. Earl. At some later date, he changed the spelling to Earle; see McCormick, 1966, p. 1.

19 For Earle's own description of his travels, see Earle, 1832, as reprinted in McCormick, 1966.

20 For a more extensive account of the life and works of Augustus Earle, see Hackforth-Jones, 1980.

21 *Correspondence*, vol. 1, pp. 550, 551; for letters relating to Covington, see *Correspondence*, vol. 1, pp. 311–15; vol. 2, pp. 194, 195, 395; vol. 4, pp. 229, 368; vol. 5, pp. 85, 163, 264; vol. 6, pp. 55, 345; vol. 7, pp. 95, 235 (corresponding to letters 206, 513, 514, 700, 1237, 1370, 1477, 1538, 1637, 1840, 2056, 2276, 2400, respectively, in *Darwin Correspondence Project* http://www.darwinproject.ac.uk); Ferguson, 1971.

22 1845 *Journal*, ch. 1, 1st paragraph. This is the first appearance of this passage; it does not appear in either the 1836 Diary or the 1839 *Journal*.

23 P. G. King (the younger)'s autobiography, in Memoirs of Governor King, Admiral King and an autobiography by Philip Gidley King, the younger, Mitchell Library C770, p. 84.

24 *Narrative*, vol. II, p. 20.

25 *Narrative*, vol. II, p. 20.

26 Information on Martens was gleaned from several sources, including Barlow, 1950; Lindsay, 1968; Gray, 1978; Dundas, 1979; and Pearce, 1979.

27 *Correspondence*, vol. 1, p. 335; letter 218 in *Darwin Correspondence Project* (http://www.darwinproject.ac.uk)

28 Literally 'a rare bird among ships', but in this context actually meaning 'a rare ship among ships'. Here, FitzRoy is using a well-known phrase (rara avis) in a somewhat distorted manner, which may have been intended as a piece of wit. We are indebted to Mrs Karen Moon for initial help with interpretation of FitzRoy's Latin, and to Mr John Sheldon for supplying the detailed information presented in this and the following paragraphs and notes.

29 In composing his line of Latin, FitzRoy aims to write a verse that scans, i.e. conforms to a particular classical measure — in this case the heroic measure, which consists of dactylic hexameters, made famous by the epic poets. In particular, FitzRoy has attempted to incorporate the well-known phrase *rara avis* from Horace's *Satires* II.2.26 (rara avis et picta pandat spectacula caudo) into his own version of an equally well-known line from Virgil's *Aeneid* II.794 and VI.702 (Par levibus ventis volucrique simillima somno). In fact, FitzRoy's line conforms only if *navibus* is scanned incorrectly as naˇvĭbuˇs; if it is scanned correctly as naˉvĭbuˇs, then the line does not conform. As explained to us by John Sheldon, if FitzRoy had written 'rara avis in navi Caroloque simillima Darwin', he would have achieved a line that does scan, and that also makes more sense, because it uses the singular form 'navi',

which then means that Martens is described as a rare one in the ship, i.e. in the *Beagle*.

30 In addition, neither *similis* nor *similior* would have conformed to the original line from Virgil.

31 In fact, like his first line, this second line does not scan because FitzRoy's memory of Latin is less than perfect: he is still giving an incorrect quantity to *navibus*. As explained by John Sheldon, FitzRoy would have achieved his dual aims (of expressing his highest admiration for Darwin and writing in the desired metre) if his second attempt had been 'est avis in navi Carolus rarissima Darwin'.

32 *Correspondence*, vol. 1, p. 354; letter 230 in *Darwin Correspondence Project* (http://www.darwinproject.ac.uk)

33 *Correspondence*, vol. 1, p. 393; letter 248 in *Darwin Correspondence Project* (http://www.darwinproject.ac.uk)

34 Mellersh, 1968, pp. 105, 297; Gribbin and Gribbin, 2003, pp. 148–58.

35 *Correspondence*, vol. 1, p. 411; letter 259 in *Darwin Correspondence Project* (http://www.darwinproject.ac.uk)

36 P. P. King Correspondence, 1824–55, pp. 55–7, Mitchell Library A3599, microfilm FM4/66.

37 Keynes, 1979. In his introduction, Keynes gives a full account of the history of Martens's sketch books.

38 *Correspondence*, vol. 1, p. 466; letter 286 in *Darwin Correspondence Project* (http://www.darwinproject.ac.uk).

39 *Correspondence*, vol. 1, pp. 471–2; letter 289 in *Darwin Correspondence Project* (http://www.darwinproject.ac.uk).

2 ARRIVAL IN SYDNEY AND A TRIP ACROSS THE BLUE MOUNTAINS

1 Details obtained from the entry for Monday 11 January 1836, p.m., in the logbook of the *Beagle*, Public Record Office, Kew, Surrey, England. ADM51/3055:6228, p. 6.

2 Eldershaw, 1973; A. Atkinson, pers. comm.

3 Letter to Susan Darwin, written in Sydney on 28 January 1836, just before the *Beagle* departed for Hobart. Other extracts from this letter are presented later in this chapter. The whole letter is reproduced in *Correspondence*, vol. 1, pp. 482–4; letter 294 in *Darwin Correspondence Project* (http://www.darwinproject.ac.uk).

4 King, P. G. (the younger), autobiography 1894, Mitchell Library FM4/6900, frame 320.

5 Memoirs of Governor King, Admiral King and an autobiography by Philip Gidley King, the younger, Mitchell Library C770, p. 88. It is interesting to note that this passage verifies that Martens' first studio was in Bridge Street and not, as has sometimes been reported, in Pitt Street or Cumberland Street.

6 *Ibid*. William Essington King was born on 8 September 1821, the third son of P. P. King and Harriet King (Mowle, 1978).

7 Memoirs of Governor King, Admiral King and an autobiography by Philip Gidley King, the younger, Mitchell Library C770, p. 82.

8 Diary of Mrs P. P. King 1835–1845, Mitchell Library B775.

9 Atkinson, A., pers. comm.; Dow, 1974.

10 Letter written to Susan Darwin on 3 September 1835, at Lima, Peru and another written to Caroline Darwin on 27 December 1835, at Bay of Islands, New Zealand. Both letters are reproduced in full in *Correspondence*, vol. 1, pp. 465–7 and 471–2, respectively; letters 286 and 289, respectively, in *Darwin Correspondence Project* (http://www.darwinproject.ac.uk).

11 Letter written to Susan Darwin on 28 January 1836, reproduced in *Correspondence*, vol. 1, pp. 482–4; letter 294 in *Darwin Correspondence Project* (http://www.darwinproject.ac.uk).

12 Soon after the commencement of the New South Wales section of Darwin's geological

diary (manuscript DAR 38.1 in the Darwin Archive at Cambridge University Library) is a page (headed 813) that contains a brief handwritten itinerary of the Bathurst trip. In contrast to the geological diary that surrounds it, the itinerary appears to have been written before the trip. Immediately following the itinerary is a list of principal landmarks along the route, showing their height above sea level. This information was probably obtained from Conrad Martens, who had recently travelled along the same road, at least as far as the Blue Mountains. It is possible that the handwritten itinerary is all that Darwin took with him to Bathurst, but it seems more likely that he would have also acquired a copy of the 1835 *Calendar*, which provides much more detail than Darwin's handwritten notes.

13 Abstract of meteorological journal, in *Narrative*, appendix to vol. II, p. 51, entry for 16 January 1836.

14 O'Shaughnessy, 1835, pp. 205–12.

15 Anon., 1836, p. 283.

16 It is not clear whether 'this country' refers to England or Australia. In fact, as Darwin would have known, many members of chain gangs had been sent to Australia as punishment for crimes committed in England, while some were serving time for offences committed in Australia.

17 In his 'Autobiography' (Barlow, 1958, p. 81), Darwin recalls how at one of the *Beagle*'s first ports of call on the voyage, it 'first dawned on me that I might perhaps write a book on the geology of the various countries visited, and this made me thrill with delight'. This explains why Darwin made such detailed geological notes throughout the voyage — he was collecting information for the geological book he had already decided to write. In the end, three geological books resulted from the voyage: *Coral Reefs* (1842), *Volcanic Islands* (1844), and *South America* (1846).

18 Malthus, 1798; 2nd rev. edn. 1803.

19 Freeman, 1978, p. 199; Clark, 1984, p. 54.

20 Freeman, 1977, p. 32; letter dated 7 July 1837, in which Darwin tells his cousin W. D. Fox that he has finished the *Journal*, and is readying it for the press; letter dated 19 Nov. 1837, in which Darwin tells J. S. Henslow that he has only one more chapter of the proofs of the *Journal* to finish. These two letters are reproduced in *Correspondence*, vol. 2, pp. 29–30 and 59–60, respectively; letters 364 and 388 in *Darwin Correspondence Project* (http://www.darwinproject.ac.uk).

21 Freeman, 1977, p. 32.

22 Newell, 1938, p. 48.

23 Newell, 1938, p. 53. Governor Bourke named it 'Lennox Bridge', a name that is still sometimes used today. This bridge is occasionally confused with another bridge by David Lennox, completed in 1839 across the Parramatta River at Parramatta. This latter bridge was named 'Lennox Bridge' by Parramatta Municipal Council in 1867 (Anon., 1950, p. 38 and p. 41).

24 Restoration commenced in the late 1970s, and the bridge was officially re-opened to traffic by the then Mayor of the Blue Mountains City, Alderman Peter Quirk, on 14 December 1982. A second re-opening ceremony was performed by the then Premier, Mr Neville Wran, on 16 September 1983. The only way to approach the bridge by car is from the town of Blaxland in the Blue Mountains: turn off the Great Western Highway at the eastern end of the Blaxland Railway Station, into Layton Avenue; follow Layton Avenue under the railway; pass straight across an intersection, and you are now at the top of Mitchell's Pass; follow Mitchell's Pass until you reach the bridge. To rejoin the highway, either retrace your steps to Blaxland, or continue down the Pass until it meets the old highway at the foot of Lapstone Hill, on the edge of Emu Plains; follow the old highway east for about 1 km until the intersection with

Russell Street; turn right into Russell Street, and follow this street for about 500 m until it joins the M4 motorway. Because of the narrowness of the Mitchell's Pass road below the bridge, it is not possible to drive up Mitchell's Pass from the Emu Plains end (Low, 1983).

25 There has been some dispute about the site of the hut. However, Cox's own account leaves no doubt as to its location: 'October 1. Began on Friday to put up the building for the second depôt. The situation is very pleasant, being on a ridge high enough in the front (which is due east) to overlook the standing timber altogether, and at the back there is a considerable quantity of ground without a tree, and a rivulet of fine spring water running through it. . . . The building for the store is 17 x 12, with 3 ft. sides, gable-ended, all weather-boards, and a door on the east end.' This clearly indicates that the hut was built on the eastern side of the creek. Cox's account is given in *Journal kept by Mr W. Cox in making a road across the Blue Mountains from Emu Plains to a new country discovered by Mr Evans to the westward, 1814,* which forms chapters VIII and IX (pp. 48–103) of *Memoirs of William Cox, J.P., Lieutenant and Paymaster of N.S.W. Corps, or 102nd Regiment, late of Clarendon, Windsor.* These two chapters were published in Mackaness, 1965, pp. 33–64. (The above entry is on p. 42.)

26 This information was obtained from Douglass, 1985, which is included as pages 63–107 in Thorp, 1985.

27 Daley, 1938, p. 70. An evergreen oak was chosen to symbolize an everlasting memorial to a native of England (Nancy Douglass, pers. comm., 1985). Douglass, 1985, p. 96, reports that on 1 November 1952, the Blue Mountains Historical Society unveiled a Weatherboard commemorative plaque attached to a large stone which was placed at the foot of the Darwin tree. By 1967, the plaque had been stolen, and the

stone was removed to the grounds of the Society's museum at 99–101 Blaxland Rd, Wentworth Falls.

28 The celebration, which included a walk following Darwin's footsteps along the last kilometre of the creek, was organized by Peter Stanbury, then Director of the Macleay Museum, University of Sydney, and was hosted by artists Venita and Reinis Zusters of Wentworth Falls. A highlight of the celebration was the unexpected return of Darwin (actor Tim Elliot), who read extracts from his *Journal* while standing on a rock above the plaque.

29 This brief summary of the history of Andrew Gardiner and his inn was taken from an informative booklet by Geoff Bates, 1981. See also Rickwood and West, 2005, especially the section on Gardiner by Peter C. Rickwood (pp. 50–55).

30 On the same day as the celebration at Wentworth Falls, the Rotary Club of Blackheath organized a re-enactment of Darwin's walk from Gardner's Inn to Govett's Leap, with local historian Edgar Penzig playing the part of Darwin. The Gardner's Inn commemorative dinner was part of these celebrations (*Blue Mountains Echo,* vol. 5, no. 52, Wed. 15 Jan. 1986, p. 4)

31 The geological statements in the three preceding paragraphs have been obtained from several sources. The origin of Darwin's marine denudation explanation was discussed in Vallance, 1975, p. 28. The estimate of the length of time since erosion commenced in the Blue Mountains was obtained from Branagan, 1985, p. 13. Darwin's change of mind about the role of erosion was described by Chorley *et al.,* 1964, p. 604. See also Pickett and Alder, 1997.

32 Williams, 1982.

33 New South Wales Department of Main Roads, 1949, which was reprinted as a booklet entitled *The Great Western Highway* in the series *Historical Roads of New South*

Wales, by the Department of Main Roads (now Roads and Traffic Authority), Sydney. Fig. 4, p. 12.

34 Winchester, 1972.

35 Hawkins, 1980, p. 10.

36 We are grateful to Diana Simpkins from the Wallerawang Library for alerting us to this event.

37 Andrew Brown later became a very influential and prosperous citizen, at one time owning approximately 78 000 hectares of land in the Castlereagh district in addition to approximately 1500 hectares at Bowenfels, near Wallerawang. A detailed and entertaining account of Andrew Brown's life can be found in Jack, 1987. Interestingly, Andrew Brown's original cottage (1824) still stands at Cooerwull, a few kilometres from the site of the Wallerawang homestead, closer to Lithgow, on Farmers Creek. Since it was Brown who was with Darwin on the evening of 19 January (when the platypus was shot), it is possible that Darwin visited Brown's cottage, or even slept in it before setting off to Bathurst on 20 January. We thank Ian Jack for this information and speculation.

38 Winchester, 1972, pp. 7–8.

39 Darwin, C. R., Fragment of an autobiography 'written August 1838', Darwin Archive, Cambridge University Library, DAR 91, p. 53.

40 Gould, 1983, pp. 284–5.

41 List of Threatened Fauna, as listed under the *Environment Protection and Biodiversity Conservation Act 1999* (EPBC Act) of the Commonwealth of Australia; for full, updated details, see the Species Profile and Threats Database, maintained under the Act, at http://www.environment.gov.au/cgi-bin/sprat/public/sprat.pl

42 *Ibid.*

43 1845 *Journal*, pp. 138, 243.

44 Blakers *et al.*, 1984, pp. 272–4, 260; Forshaw and Cooper, 1981, pp. 185–90, 197–202, 144–8; Forshaw and Cooper,

1978, pp. 234–6, 238–40, 213–16. We are grateful to E. Arnold, Lithgow, and J. Dark, Hazelbrook, for advice on which species of parrots and other birds are likely to be amongst those mentioned by Darwin.

45 Blakers *et al.*, 1984, p. 249; Forshaw and Cooper, 1981, pp. 100–5; Forshaw and Cooper, 1978, pp. 129–31.

46 Blakers *et al.*, 1984, p. 644; Goodwin, 1977, pp. 64, 74–8, 110–12.

47 Blakers *et al.*, 1984, p. 640–1; Goodwin, 1977, pp. 173–81.

48 John Wrigley states that 'the common name of the she-oak is said to be derived from the inferior oak-like timber, but in these days of equality of sexes, the author stakes no claim to its origins' (Wrigley and Fagg, 1983, p. 327).

49 When the animal was first described by Shaw, 1799, the name given was *Platypus anatinus*, from the Greek *platys* 'flat' and *pous* 'foot' and Latin *anatinus* 'duck-like', respectively. A year later, Blumenbach, 1800, called it *Ornithorhynchus paradoxus* – the name used by Darwin. Later, the original species name was preferred, but the original generic name could not be used, having already been allocated to a genus of beetles. The present name, which was used as early as 1846 by Waterhouse, is a combination of the others: *Ornithorhynchus anatinus*. (Dawson 1983, p. 9; Gould, 1983, p. 2). See also Moyal, 2001.

50 Collins, 1802, ch. VI.

51 Gould, J., 1983, p. 2.

52 Dawson, 1983, p. 18; Ride and Fry, 1970, p. 194; Hyett and Shaw, 1980, p. 24.

53 Gould, J., 1983, p. 4.

54 Fleay, 1980, p. 46; Dawson, 1983, p.18.

55 Kirby and Spence, 1815–26. On board the *Beagle*, Darwin had vol. 1 of the 3rd edn, vol. 2 of the 2nd edn, and vols 3 and 4 of the 1st edn. All were heavily annotated (*Correspondence*, vol. 1, p. 561). Since it is most unlikely that he would have carried these

books with him on the trip to Bathurst, the fact that he quoted the actual page number in the middle of an original sentence (it is not an insertion) in the manuscript Diary is evidence that the Diary was written up after he had returned to the *Beagle*.

56 For those who may wish to make a detailed comparison, the complete ant-lion passage in the 1839 *Journal* is:

A little time before this I had been lying on a sunny bank, and was reflecting on the strange character of the animals of this country as compared with the rest of the world. An unbeliever in every thing beyond his own reason might exclaim, "Two distinct Creators must have been at work: their object, however, has been the same, and certainly the end in each case is complete." While thus thinking, I observed the hollow conical pitfall of the lion-ant: first a fly fell down the treacherous slope and immediately disappeared; then came a large but unwary ant; its struggles to escape being very violent, those curious little jets of sand, described by Kirby* as being flirted by the insects tail, were promptly directed against the expected victim. But the ant enjoyed a better fate than the fly, and escaped the fatal jaws which lay concealed at the base of the conical hollow. There can be no doubt but that this predacious larva belongs to the same genus with the European kind, though to a different species. Now what would the sceptic say to this? Would any two workmen ever have hit upon so beautiful, so simple, and yet so artificial a contrivance? It cannot be thought so: one Hand has surely worked throughout the universe.

*Kirby's Entomology, vol. i., p. 425. The Australian pitfall is only about half the size of the one made by the European species.

57 Darwin, 1887: vol. I, pp. 83–4; vol. II, ch. 1. Both the 1842 and 1844 essays were published in Darwin, 1909, and have been republished in Barrett and Freeman, 1986, vol. 10.

58 The following words appear in a footnote to the word 'contracted', which is the last word in the section on the platypus:

I was interested by finding here the hollow conical pitfall of the lion-ant, or some other insect: first a fly fell down the treacherous slope and immediately disappeared; then came a large but unwary ant; its struggles to escape being very violent, those curious little jets of sand, described by Kirby and Spence (Entomol., vol. i., p. 425) as being flirted by the insect's tail, were promptly directed against the expected victim. But the ant enjoyed a better fate than the fly, and escaped the fatal jaws which lay concealed at the base of the conical hollow. This Australian pit-fall was only about half the size of that made by the European lion-ant.

59 Freeman, 1978, pp. 177–8.

60 In a letter to the publisher John Murray on 6 June 1845, Darwin said that he had much material that he wished to add about the Fuegians, and much to condense and rewrite in the scientific parts. In another letter to Murray, on 23 June 1845, Darwin explained that he wanted the word 'Journal' to stand alone in the title, so that the scientific aspects of the book would be less prominent. In a letter written before 9 July 1845, to Ernst Dieffenbach, who had recently translated the first edition of the *Journal* into German, Darwin stated 'I have largely condensed, corrected & added to the Second English Edition, & I am sure have considerably improved & popularized it.' (*Correspondence*, vol. 3, pp. 204, 205, 215; letters 876, 878, and 888, respectively, in *Darwin Correspondence Project* (http://www.darwinproject.ac.uk).

61 The details of print runs by John Murray were kindly made available by Ms V.

Murray of John Murray (Publishers) Ltd. A list of the translations is provided by Freeman, 1978, p. 178, and details of all editions are given by Freeman, 1977, pp. 31–54. Given the remarkable success of the book, Darwin must have lived to regret having foregone any income from royalties by selling the copyright for the second edition to Murray for only £150 (Freeman, 1978, p. 177; Freeman, 1977, pp. 34–5).

62 In a letter to his cousin W. D. Fox, written in April 1844, Darwin tells Fox that a strange book (*Vestiges*) has appeared, and that some people think that he (Darwin) has written it. (*Correspondence*, vol. 3, p. 180; letter 859 in *Darwin Correspondence Project* (http://www.darwinproject.ac.uk).

63 As explained in the Preface, the geological diary is a different manuscript source from the personal Diary. This is the first time in the present book that the former has been quoted.

64 Anon., 1836, pp. 328 and 334.

65 Anon., 1836, pp. 360–2.

66 *Correspondence*, vol. 1, p. 481; letter 293 in *Darwin Correspondence Project* (http://www.darwinproject.ac.uk).

67 Included among the houses that Darwin may have seen are Blackdown (Thomas and Elizabeth Hawkins), Kelsoville (Richard Cousins), Mt Tamar or Wonalabee (McPhillamy family), Keloshiel (George Ranken), Strath (Major General Stewart), Littlebourne (Francis Lord), Macquarie (Lawson family), Springdale or Abbotsford, Alloway Bank (Captain John Piper), and Woolstone. All of these houses were located within 8 kilometres of Bathurst. (The above list of houses and owners, together with details of their locations, was kindly supplied by Mr Theo Barker and Mrs Carol Churches, Bathurst District Historical Society.)

68 Reed, 1978, p. 37. It is interesting to note that one of Holy Trinity's claims to fame is that it was the first church to be consecrated in Australia. Although it is by no means the oldest church in Australia, none of the others had been consecrated because, until 1836, there was no bishop in the colony to do the consecrating. Soon after Holy Trinity's foundation stone had been laid in February 1834, the person who had performed the ceremony, Archdeacon Broughton, returned to England. Eighteen months later, he returned to Australia as the colony's first bishop, and the first church he consecrated was the recently-completed Holy Trinity, on 3 December 1836.

69 Adams, 1952, p. 34; as quoted by Reed, 1978, p. 38.

3 RETURN FROM BATHURST AND IMPRESSIONS OF SYDNEY

1 The Lockyer Papers, compiled by Nicholas Lockyer, 1918, Mitchell Library MSS2513, pp. 72ff.

2 The routes of Mitchell's and Lockyer's roads (and also Cox's road – the first to be built, in 1815) are shown in the map facing p. 15 in *Report Upon the Progress Made in Roads and in the Construction of Public Works in New South Wales from the Year 1827 to June 1855* by Colonel Sir T. L. Mitchell, Surveyor General, 1856, Mitchell Library A331. This map is redrawn in Newell, 1938, fig. 9, p. 47; and in *Main Roads* 15(1), 1949, 6–15, fig. 4, p. 12; and in an undated booklet entitled *The Great Western Highway* in the series *Historical Roads of New South Wales* (Dept of Main Roads, Sydney). Lockyer's road is not labelled as such, but is referred to as 'the road cleared by mistake in 1829'. As indicated by the itinerary which is reproduced in the present book, Lockyer's road and Mitchell's road cross each other at several points near South Bowenfels and Hassan's Walls. A detailed map showing these intersections is given in McKenzie, 1966, map A, p. 6.

3 The itinerary in Anon., 1835b, records the distance from Sydney to Weatherboard as 58½ miles (94 kilometres), and from Sydney to Bathurst as 114 miles (183 kilometres). The total distance from Sydney to Bathurst and back to Weatherboard would therefore be 114 + (114 − 58½) = 169½ miles (272 kilometres). This is a conservative estimate, because it does not allow for the detour to Wallerawang.

4 *Correspondence*, vol. 1, p. 561. The paper is King and Broderip, 1832–4, which is reprinted in the *Narrative*, vol. I, pp. 545–56 (actually 545–60, because page numbers 556–9 are repeated in this volume of the *Narrative*). Also included in volume I of the *Narrative* are catalogues of mammals (pp. 529–31) and birds (pp. 532–44), and extracts from the *Beagle*'s game book (pp. 586–7), all pertaining to the *Beagle*'s first surveying voyage. When Darwin visited King at Dunheved, the catalogues would have been in preparation, and were undoubtedly discussed at great length.

5 Pike, 1967, pp. 61–4; Walsh, 1967, p. 59.

6 For a very thorough heritage assessment, see Anon., 2005; for its listing on the Australian Heritage Database, see http://www.environment.gov.au/cgi-bin/ahdb/search.pl. We thank Mary Casey, the author of the heritage assessment, for advising us of the future plans for the site.

7 Gregson, 1907.

8 The entry for Tuesday 26 January, p.m., in the logbook of the *Beagle*, Public Record Office, Kew, Surrey, England, ADM51/3055:6228, p. 8.

9 Anon., 1950, pp. 39–40, reprinted in a booklet entitled *Bridge Building in New South Wales, 1788–1938*, by NSW Department of Main Roads, Sydney, n.d.; Herman, 1954, pp. 161–4. The Lansdowne Bridge is still in use today, carrying all Sydney-bound Hume Highway traffic over Prospect Creek.

10 Pike, 1967, pp. 147–9; Verge, 1962, pp. 72–4; Wilson, 1917–19, p. 387; Clifford,

1961. The reference to the house being under construction occurs in *The Colonist*, Thurs. 6 Aug. 1835, vol. I, no. 32, p. 254, 2nd col.

11 Clifford, B., 1961, and photographs in the possession of the Archivist, University of New South Wales.

12 de Falbe, 1860–94; The recollections were published as de Falbe, 1954a, 1954b, 1988. We can conjecture how Vineyard was run from Barca, 1979. This little book presents a long letter sent from England to Maria Macarthur by a woman thought to be her godmother. It provides menus for a variety of meals from 'a sociable sandwich' to 'a grand dance', and gives innumerable hints on rules of etiquette and how to handle servants.

13 Anon., n.d.(b); Pike, 1967, p. 597 (Wickham), and Pike, 1974, p. 29 (King).

14 de Falbe, 1860–94; Stokes, 1846, vol. I, pp. 241–57, vol. II, pp. 243–50, 413–17, 497–9; Marshall, 1970, pp. 82–113.

15 The nuns moved again in 1988, to Jamberoo, south of Sydney, where their life was featured in the 2007 ABC TV documentary *The Abbey*.

16 Information on Subiaco was kindly provided by Sister Marie Therese, Benedictine Priory, then at West Pennant Hills. The quotation is from de Falbe, 1954a, p. 24. Other sources are Roxburgh, 1974; Broadbent and Dupain, 1978, pp. 22–7; Wilson, 1924; Walsh, 1967, p. 51; Cole, 1980 (reprinted as ch. 15 in Cole, 1983); Anon., 1967; Pickard, 1976, p. 4, and reply by Lucas, 1976, p. 2; Mitchell Library newspaper clippings, Sydney — Residences — Vineyard, Subiaco (*Sydney Morning Herald* 26.11.60 p. 13, 30.11.60, 27.5.61, 19.6.61, 21.7.61, and *Daily Mirror* 26.1.61). The house was bought and demolished by Rheem Australia, who sold the site to Prudential Assurance in 1981. The Prudential Industrial Estate was opened in 1985. Fortunately, some of the fabric of

the house was saved. The fireplaces are in Macquarie Fields House: the inner doorway, fanlight and sidelights form the entrance to the museum of the King's School at Parramatta; and the fine colonnade of Doric columns has been re-erected beside the Village Green Oval at the University of New South Wales in Kensington. The cantilevered stone staircase has been reported as being donated to the University of New South Wales, but the University has no record of this (Mr L. Dillon, University Archivist, pers. comm., January 1987).

17 In Darwin's time, there were three ferries in the colony: the Emu Ferry (which he had used twice), the Bedlam Point Ferry over Parramatta River, and Wiseman's Ferry over the Hawkesbury River at Wiseman's Reach (O'Shaughnessy, 1835, p. 118).

18 The fresh beef and vegetables arrived quite regularly. On isolated occasions, other goods were delivered. These included 5400 lb. bread, 392 lb. raisins, 784 lb. sugar, 334 lb. chocolate, 80 lb. tea, 6 bushells peas, 300 lb. soap, 588 lb. flour, plus various stores for the boatswain and carpenter, including rope, canvas, twine, paint, turpentine, salted hides, tanned leather, tallow, illuminator, whiting, fish hooks, cedar, and linseed oil (logbook of the *Beagle*, Public Record Office, Kew, Richmond, Surrey, England, ADM51/3055: 6228). 'Employed variously on Ship's duty' is the most common entry in the logbook, and is often the only entry for a shift. The officers' life was not all drudgery, however. On the same day that Darwin arrived back at the *Beagle*, some of the officers had attended a fete in the grounds of what was to become Alexander Macleay's Elizabeth Bay House – then under construction. Somewhat surprisingly (given Alexander Macleay's long-term secretary-ship of the Linnean Society in London prior to his move to Sydney), Darwin makes no mention of any meeting with Alexander

or any of his family during his visit to Sydney. After returning to London, Darwin spent considerable time with Alexander's son, William Sharp Macleay, until that Macleay's departure to live in Sydney in March 1839. We thank Ashley Hay for advising us of the fete, which is described in Evely, 2003.

19 The autobiography of P. G. King (the younger), in Memoirs of Governor King, Admiral King and an autobiography by Philip Gidley King, the younger, Mitchell Library C770, p. 88.

20 From a comparison of FitzRoy's estimate of the difference in longitude between Macquarie Fort and the Parramatta Observatory (52 secs on the chronometers), with a present-day estimation using Google Earth (0.21975°), the present authors have calculated that FitzRoy's estimate was around 250 metres too short.

21 The evidence for most of these statements is presented in the following paragraphs of the text. Evidence on the number of chronometers is given in *Narrative*, vol. II, appendix, p. 338.

22 This date has been determined indirectly in the following manner. Judging from the day shown on Macarthur's letter to P. P. King (which is otherwise undated), the party must have been held on a Monday, and there were only two Mondays during the *Beagle*'s stay in Sydney: the 18th and the 25th. Since Macarthur presumably had some prior knowledge that King was intending to bring Darwin to Vineyard for lunch one day during the first half of the week beginning the 25th (depending on when Darwin arrived at Dunheved after his trip to Bathurst), it is most unlikely that Macarthur would have written a letter to King on Tuesday 26, advising that he, Macarthur, was setting out that afternoon to make the (24 kilometre) journey to Dunheved to visit King. Furthermore, it is most unlikely that Macarthur could have made the trip to

Dunheved on the Tuesday afternoon, and yet have been back at Vineyard the next day, to entertain Darwin at lunch. Also, in his letter to P. G. King, dated Monday 25 January (see p. 103), FitzRoy states that he is unable to come to Parramatta as he had intended. Since FitzRoy was obviously at a party at Government House in Parramatta on a Monday night, FitzRoy's letter to P. G. King must be referring to a second visit to Parramatta that he had intended to make about a week after the party. For these reasons, it is far more likely that Macarthur's letter was written on the previous Tuesday, i.e. on 19 January. Consequently, the most likely date for the party in Government House at Parramatta is Monday 18 January. (See letter on pp. 101–2)

23 The postscript in the letter refers to General Ralph Darling, who had been governor of the colony from December 1825 to October 1831. He was knighted by the King on 2 September 1835, the day after a Select Committee of the House of Commons had cleared him of accusations of wrongdoing during his term as governor. The news had just reached the colony.

24 A letter sent to P. P. King from H. H. Macarthur, King Papers, vol. 1, 1799–1829, pp. 453–6, Mitchell Library A1976.

25 A letter sent to P. G. King from Captain FitzRoy, King Papers, vol. 2, pp. 96–8, Mitchell Library A1977.

26 P. G. King, *loc. cit.*

27 *Correspondence*, vol. 1, pp. 482–4; letter 294 in *Darwin Correspondence Project* (http://www.darwinproject.ac.uk).

28 1845 *Journal*, ch. IX.

29 Phillip, 1789. This passage and the poem were quoted by FitzRoy in *Narrative*, vol. II, pp. 621–2.

30 *Narrative*, vol. II, p. 622.

31 *Correspondence*, vol. 1, pp. 484–6; letter 295 in *Darwin Correspondence Project* (http://www.darwinproject.ac.uk).

32 Darwin's ideas on missionaries were influ-enced by FitzRoy's strongly held beliefs. A detailed account of the two men's thoughts at that time is presented in FitzRoy and Darwin, 1836 (reprinted in Barrett, 1977, vol. 1, pp. 19–37).

33 For full details of the Sydney specimens, see Nicholas, 2008.

34 Waterhouse, 1838–9, number 4, pp. 67–8; see also Nicholas, 2008. For information on the extinct status of this species, see http://www.environment.gov.au/cgi-bin/sprat/public/publicspecies.pl?taxon_id=89 and http://www.iucnredlist.org/search/details.php/18551/all

35 The identity of this species is not clear. From the list of Darwin's Australian specimens (Lea, 1926), the most likely possibility is *Novius bellus*, a red ladybird beetle with black markings. We thank Dr D. S. (Woody) Horning for this suggestion.

36 The insects collected in Sydney are listed in Smith, 1987, pp. 99–100. Dr Horning advises that *Idiocephala darwinii* is now called *Loxopleurus darwinii*. See also Nicholas, 2008.

37 The original letters have been lost, but relevant extracts from them were published in Darwin, 1912.

38 A letter sent to Major T. L. Mitchell from Captain FitzRoy. Mitchell, Sir T. L., Papers 1830–1839, vol. 3, pp. 217–20, Mitchell Library CYA292.

39 Conrad Martens' account book, p. 3, records the sale of a painting called *View at Moorea* to 'Rob. FitzRoy Esq Capt RN', for 2 guineas, on 28 January 1836. Dixson Library CYMS142.

40 Included among the 'Shore attractions' was of course, Philip King's cousin Elizabeth. As we have already seen, she was one of the 'very nice looking young ladies' whom Darwin met during his lunch at Vineyard. In 1843 she and Philip King were married.

41 P. G. King autobiography, *op. cit.*, p. 88.

42 Reproduced with permission from Stanbury, 1977, pp. 337–8.

4   HOBART AND ENVIRONS

1   The history of the Iron Pot Light involves an interesting coincidence with another former member of the *Beagle*'s crew, namely Augustus Earle. The details of this coincidence are given in the second paragraph of note 5 below.

2   Basalt and dolerite both result from the cooling of molten material (magma) that has originated beneath the earth's surface. The difference is that basalt forms when a particular type of magma is extruded at the surface as lava, whereas dolerite is one possible result when the same type of magma cools and solidifies beneath the surface. Darwin's geological observations in Tasmania, including this mistake, have been fully discussed in Banks, 1971 and Banks and Leaman, 1999.

3   Anon., n.d.(a).

4   Robertson, 1970, vol. 1, ch. 5.

5   Earle's last departure from Sydney on 12 October 1828, on board HMS *Rainbow*, is documented in Hackforth-Jones, 1980, p. 149. The *Rainbow*'s visit to Hobart from 19 October to 11 November 1828 is documented in Nicholson, 1983, pp. 149–50. We are grateful to Professor R. G. Beidelman for bringing Earle's second Hobart visit to our attention. From Nicholson's book, it is evident that after only four days in Hobart, the *Rainbow* was commissioned for a short cruise to the mouth of the Derwent River, for the purpose of determining the best site for a lighthouse. With the Lieutenant Governor, Colonel George Arthur, on board, the *Rainbow* was engaged on this task from 23 October to 2 November 1828. The site chosen was Iron Pot Island, off the southern tip of South Arm. A lighthouse designed by John Lee Archer was erected on the island in 1833 (Anon., 1983, p. 16; See also Reid, 1988). Three years later, it was this lighthouse that was sketched by Darwin's servant Syms Covington as the *Beagle*

entered the mouth of the Derwent River. The Iron Pot Light still stands today, and is the oldest lighthouse structure in Australia. After returning from its short cruise, the *Rainbow* stayed in port for a further nine days, before departing on 11 November bound for Madras in India, via the Caroline Islands, Guam, Manila and Singapore. Assuming that Earle did not go on the short cruise, he spent a total of twenty-two days in Hobart during his second and last visit. In his panorama, Earle shows three ships: the *Rainbow*, the *Mermaid*, and the *Cyprus*, the latter being the ship on which he had originally sailed from Hobart to Sydney in 1825. Nicholson's book has no record of the *Rainbow* being in Hobart in 1825, but it does show that in that year, the *Mermaid* was in Hobart from 21 to 28 April, and the *Cyprus* was in port from 21 April to 6 May. Three years later, the *Mermaid* was in Hobart for much of the time between 2 October 1828 and 4 April 1829, and the *Cyprus* arrived again in Hobart on 13 September 1828. Although there is no record of when the *Cyprus* departed, it was not included in a list of ships in harbour on 8 October 1828, but both the *Cyprus* and the *Mermaid* were listed as being in harbour on 1 November of the same year. It is possible, therefore, that the *Cyprus* and the *Mermaid* were sketched in either 1825 or 1828. The *Rainbow* must have been added in 1828.

6   The account given in the following paragraphs has been obtained from Turnbull, 1948; Ryan, 1981; Robson, 1983; and Robson, 1985.

7   Ryan, 1981.

8   In the 1845 *Journal*, Darwin shortened the list of interesting points by excluding the first two. He also changed the order of the remaining points, slightly decreased his estimate of the age of the fossils, and made several other alterations. The 1845 sentence is: 'The main points of interest consist, first in some highly fossiliferous strata,

belonging to the Devonian or Carboniferous period; secondly, in proofs of a late small rise of the land; and lastly, in a solitary and superficial patch of yellowish limestone or travertine, which contains numerous impressions of leaves of trees, together with landshells, not now existing.' The change in the estimate of the age of the fossils was made on the advice of W. Lonsdale, who described six of the specimens collected by Darwin, on pp. 161–9 of *Volcanic Islands*.

9   Banks, 1971; Banks and Leaman, 1999.

10 Banks has noted that the 'place of contact' between Triassic sandstone and greenstone (dolerite) that Darwin observed in present-day Queens Domain can be seen on the shoreline near the Naval Depot.

11 The date for this excursion can be deduced from the DAR40 manuscript.

12 O'May, 1959, ch. 6, pp. 61–3. See also Norman, 1938, p. 146.

13 The manuscript has 'Tuesday 10th', but this must be incorrect, because 10 February 1836 was a Wednesday. Since it is clear from Darwin's Diary that Wednesday 10 was taken up with the unsuccessful attempt to climb Mt Wellington, the correct entry must be 'Tuesday 9th'.

14 Dr Banks advises that the 'globular concentric structure' is a reflection of onion-skin weathering – a common form of weathering of dolerite.

15 We thank Dr Banks for informing us of this gazetting, which was arranged in preparation for commemorating the 200th anniversary of Darwin's birth on 12 February 2009. The age of the volcanic rocks has been estimated to be approximately 26.5 million years (Sutherland and Wellman, 1986).

16 Dr Banks has noted that Darwin's estimate of the height of Mt Wellington is about 320 metres too low. The correct height is 1270 metres (4167 ft).

17 Covington, 1831–6, p. 6.

18 Pike, 1966, pp. 410–11; Robson, 1983), p. 181.

19 Robert Brown was naturalist with Matthew Flinders on board the *Investigator*. He spent several months in Tasmania during the summer of 1803–4, when the *Investigator* was undergoing repairs in Sydney. Among other things, he collected fossils in Tasmania, including the first Australian fossil to be described. Later he became a leading British botanist. From 1810 to 1820, he was librarian to Sir Joseph Banks, and between 1827 and 1858 he was the keeper of the botanical collections at the British Museum (*Correspondence*, vol. 1, p. 616). He was also the discoverer of what is now called Brownian motion. (The information on Brown's Tasmanian activities was kindly supplied by Dr Banks.)

20 Robertson, 1970, vol. 1, p. 174; Anon., 1983, p. 33; Dupain and Herman, 1963, p. 145; Rowntree, 1951.

21 Bedford, 1954; Nairn, 1976, pp. 180–7; Anon., 1983, p. 48; Anon., 1964, p. 125. Now known as Stephenville, the house forms part of St Michael's Collegiate School.

22 The *Beagle*'s official meteorological record says the weather was *bcq* (blue skies, passing clouds, and squally). However, the log books record it as being *g* (gloomy: dark weather) as well.

23 *Correspondence*, vol. 1, pp. 489–91; letter 298 in *Darwin Correspondence Project* (http://www.darwinproject.ac.uk).

24 *Correspondence*, vol. 1, pp. 491–3; letter 299 in *Darwin Correspondence Project* (http://www.darwinproject.ac.uk).

25 We are grateful to Dr D. S. (Woody) Horning for this information about the carab beetles mentioned by Darwin. Some of Dr Horning's information was obtained from Linssen, 1959. The words 'Badister, which was not cephalotes' imply that *cephalotes* was a species of *Badister*, but Dr Horning can find no evidence of this.

The situation is complicated by the fact that the genus *Cephalotes* has since been synonymised under *Broscus*, and that a member of this genus called *B. cephalotes* occurs in England.

26 See note 31 in Chapter 3.

27 Frankland, 1836.

28 Cogger, 1983, pp. 387–8. We are grateful to Dr Cogger for verifying our interpretation of his published account of the oak skink.

29 *Zoology*, part V, p. 30. This passage is a slightly edited version of the original notes in the zoological diary (Darwin Archive, Cambridge University Library), DAR 31.2, p. 346.

30 Zoological diary, Darwin Archive, Cambridge University Library, DAR 31.2, p. 347.

31 *Ibid.*, p. 348. The words in double quotes refer to colour standards in Syme, 1821, a copy of which was in the *Beagle*'s library.

32 *Ibid.* The modern equivalent to Darwin's term 'ovoviparous' is 'ovoviviparous'. Both terms refer to live-bearing of offspring which have developed without a placenta, cf. 'oviparous', meaning egg-laying.

33 This interpretation of Darwin's comments has been verified by Dr Cogger.

34 Zoological diary, DAR 31.2, p. 363.

35 *Ibid.*, p. 364.

36 *Ibid.*

37 *Ibid.*, p. 365.

38 Smith, 1987, pp. 97–9.

39 Bornemissza, 1983. The quoted passage is from 1839 *Journal*, p. 583.

40 'Mr. Duff' is probably Lieutenant William H. Duff (Hügel, 1994, pp. 62, 63).

5 KING GEORGE SOUND AND FAREWELL TO AUSTRALIA

1 Many authors still use the name King George's Sound, but the official name is King George Sound (Australia Division of National Mapping, Department of Minerals and Energy, Canberra [Author], *Australia 1:250,000 Map Series Gazetteer* [Australian Government Publishing Service, Canberra, 1975], p. 469); see also http://www.ga.gov.au/map/names/. The historical facts in this chapter have been obtained primarily from Garden, 1977, which is an official history of Albany, commissioned by the Council of the Town of Albany to celebrate the town's sesquicentenary.

2 The initial instruction from the Secretary of State, Lord Bathurst, proposed that the settlement be established at Shark Bay, about 1100 kilometres north of King George Sound. However, in a second letter written soon after the first, this was changed to King George Sound, because Lord Bathurst had heard that the area around Shark Bay was too barren (Garden, 1977, p. 14).

3 In 1832, the name of the settlement was officially changed to Albany, but the name King George Sound remained in general use for many decades.

4 *Narrative*, vol. II, p. 625.

5 *Ibid.*

6 Johnson, 1984, p. 18; Chessell, 2005. See also Hügel, 1994, pp. 76–95.

7 The new house was built adjacent to the wattle-and-daub building that Darwin and FitzRoy visited, adjoining it at one end. In 1870, the latter building was destroyed by fire. See Johnson, 1984; Lukis, n.d.; and Garden, 1977.

8 *Narrative*, vol. II, p. 626.

9 Armstrong, 1985.

10 In preparing the Diary for publication, Darwin rewrote this paragraph, but did not make any significant changes to the meaning:

One day I accompanied Captain FitzRoy to Bald Head; the place mentioned by so many navigators, where some imagined they saw coral, and others petrified trees, standing in the position in which they grew.

*According to our view, the rock was formed
by the wind heaping up calcareous sand,
during which process, branches and roots
of trees, and land-shells were enclosed;
the mass being afterward consolidated by
the percolation of rain-water. When the
wood had decayed, lime was washed into
the cylindrical cavities, and became hard,
sometimes even like that in a stalactite.
The weather is now wearing away the
softer rock, and in consequence the casts
of roots and branches project above the
surface: their resemblance to the stumps
of a dead shrubbery was so exact, that,
before touching them, we were sometimes
at a loss to know which were composed of
wood, and which of calcareous matter.*

In the 1845 *Journal*, the consolidation was
attributed to percolation of calcareous matter rather than rain-water, and the cavities
were said to be filled with a 'hard pseudo-
stalactitical stone' rather than lime. The
complete 1845 *Journal* version is:

*One day I accompanied Captain Fitz Roy
to Bald Head; the place mentioned by so
many navigators, where some imagined
that they saw corals, and others that they
saw petrified trees, standing in the position
in which they had grown. According to
our view, the beds have been formed by
the wind having heaped up fine sand,
composed of minute rounded particles of
shells and corals, during which process
branches and roots of trees, together with
many land-shells, became enclosed. The
whole then became consolidated by the
percolation of calcareous matter; and the
cylindrical cavities left by the decaying of
the wood, were thus filled up with a hard
pseudo-stalactitical stone. The weather is
now wearing away the softer parts, and
in consequence the hard casts of the roots
and branches of the trees project above*

*the surface, and, in a singularly deceptive
manner, resemble the stumps of a dead
thicket.*

11 Semeniuk and Meagher, 1981.

12 The quotation from Darwin was given by
George R. Waterhouse during the description of this new species, on p. 66 of *Zoology,
Part II, Living Mammals*. Waterhouse
called the new species *Mus fuscipes*, but it
is now recognized as belonging to the rat
genus (Strahan, 1983, pp. 440–5); see also
the Australian Biological Resources Study
(ABRS) Fauna Online website: http://www.
environment.gov.au/biodiversity/abrs/
online-resources/fauna/index.html

13 Strahan, 1983. The different geographic
populations were originally regarded as
different subspecies, but they freely inter-
breed (Horner and Taylor, 1965) and hence
are now regarded as a single species. Since
the bush rat is today one of Australia's most
common native mammals, it may seem
surprising that it was not discovered until
1836. We are grateful to Dr Dan Lunney of
the National Parks and Wildlife Service in
Sydney for explaining that this is most likely
due to the fact that although common, it is
very timid and hence is rarely seen. Also,
since it looks so much like a ship's rat, it
was probably of much less interest to early
naturalists than native marsupials.

14 Darwin's specimen was described by
Thomas Bell on pp. 33–4 of *Zoology, Part
V, Reptiles*. When the species was first described, by Tschudi, 1838, it was the only
member of the *Crinia* genus. Bell, who
attributed the species' original discovery
to Quoy and Gaimard, disagreed with
Tschudi's classification, and renamed it
Cystignathus georgianus. It is now thought
that Tschudi was correct, and so his original
name is also the present-day name, and it is
once more the type species of a monotypic
genus (Cogger, 1983, pp. 42–3); see also

the Australian Biological Resources Study (ABRS) Fauna Online website: http://www. environment.gov.au/biodiversity/abrs/ online-resources/fauna/index.html

15 *Narrative*, vol. II, p. 628.

16 The scientific and common names given in the text are the current names, as listed in the Australian Biological Resources Study (ABRS) Fauna Online website: http:// www.environment.gov.au/biodiversity/ abrs/online-resources/fauna/index.html. In arriving at these current names, we were guided by the list of current names provided by Dr J.R. Paxton when the first edition was being prepared. Only one of the names has remained unchanged since part IV (Fish) of the *Zoology* was published in 1840–2. The scientific names used in *Zoology* are given in the following list.

SCIENTIFIC NAME

CURRENT (2008)	ZOOLOGY (1840–2)
Pelates sexlineatus	*Helotes otolineatus*
Leviprora inops	*Platycephalus inops*
Trachurus declivis	*Caranx declivis*
Nelusetta ayraudi	*Aleuteres velutinus*
Arripis georgianus	*Arripis georgianus*
Pseudocaranx dentex	*Caranx georgianus*
Aldrichetta forsteri	*Dajaus diemensis*
Pseudorhombus jenynsii	*Platessa sp.*
Acanthaluteres spilomelanurus	*Aleuteres maculosus*
Family Scorpaenidae	*Apistus sp.*

17 The common name in bold is the preferred common name, as listed in the Australian Biological Resources Study (ABRS) Fauna Online website: http://www.environment.

gov.au/biodiversity/abrs/online-resources/ fauna/index.html

18 See entries for *Pelates sexlineatus* and *Nelusetta ayraudi* in the Australian Biological Resources Study (ABRS) Fauna Online database at: http://www. environment.gov.au/biodiversity/abrs/ online-resources/fauna/index.html

19 Catalogues of specimens, Darwin Archive, Cambridge University Library, DAR 29.1, p. 11.

20 Smith, 1987, pp. 100–1; for the official current information on these species, see the relevant entries in the Australian Biological Resources Study (ABRS) Fauna Online database at: http://www.environment.gov. au/biodiversity/abrs/online-resources/ fauna/index.html

21 Zoological diary, Darwin Archive, Cambridge University Library, DAR 31.2, p. 365.

22 *Ibid.*, p. 366.

23 Darwin, 1844b, p. 246; reprinted in Barrett, 1977, vol. 1, pp. 182–93. The results of Darwin's bisections of the Hobart specimens were included in chapter 2 of the 1839 and 1845 *Journals* (p. 31 and p. 27, respectively). The account corresponds closely to the extracts from the zoological diary reproduced in the present book. We are grateful to Dr Tony Underwood for verifying our interpretation of Darwin's observations on the Tasmanian planaria.

24 Journal of Sir Richard Spencer, Battye Library 1200A, pp. 3a, 3b.

25 In Beard *et al.* 2000, the number of native species is given as 5710, of which 4524 or 79.2% are endemic; for the latest information, see FloraBase (the Western Australian Flora database) at http:// florabase.calm.wa.gov.au/

26 Erickson *et al.*, 1973, pp. 60–9, 173–5; Mullins and Baglin, 1978; Wheeler *et al.*, 2002; see also entries for individual

species in FloraBase (the Western Australian Flora database) at http://flora base.calm.wa.gov.au/

6 POSTSCRIPT

1 Information on Covington was obtained from Ferguson, 1971; de Beer, 1959; and Anon., 1902. The whole of Covington's rather brief *Beagle* diary, together with much background information on his life, is available at http://www.asap.unimelb. edu.au/bsparcs/covingto/contents.htm. For a lively fictional account of Covington, see McDonald, 1998.

2 Linnean Society of London, Reel 1, Arch-ives: Correspondence, 1805–1870 (within Macleay Correspondence), Mitchell Library FM4/2699. The transcript reproduced here is taken from *Correspondence*, vol. 2, pp. 194–5; also available as letter 513 in *Darwin Correspondence Project* (http://www.darwinproject.ac.uk).

3 The letter is reproduced in *Correspondence*, vol. 2, p. 195; also available as letter 514 in *Darwin Correspondence Project* (http://www.darwinproject.ac.uk).

4 Sir Thomas Livingstone Mitchell's papers, volume VI, A295-1, misc. pp. 1–3, Mitchell Library CYReel A295/1. The transcript reproduced here is taken from *Corres-pondence*, vol. 2, pp. 195–6; also available as letter 515 in *Darwin Correspondence Project* (http://www.darwinproject.ac.uk).

5 Dietz, 1978.

6 Covington's death certificate, dated 19 February 1861, records that he had lived '21 years in this colony' (Ferguson, 1971, p. 7).

7 *Sydney Mail*, vol. 38, no. 1257, 9 August 1884, pp. 254–5. It appears that these letters were in the possession of the Linnean Society of New South Wales for many years, but, with the exception of one letter that is still in the possession of Covington's descendents, they cannot now be located.

The transcriptions presented here were also published in de Beer, 1959.

8 In 1842, the year in which Darwin and his family moved into Down House, the Government officially changed the spelling of the village's name from Down to Downe, so as to avoid confusion with County Down in Ireland. Darwin, however, refused to accept this change, and stuck doggedly to the original spelling for both his house and the village (Clark, 1984, p. 71).

9 Freeman, 1977, pp. 194–6. (Freeman's list includes twenty-four papers published up to the end of 1843. However, the first two of these have been excluded from the count because one was a letter on missionaries written primarily by FitzRoy, and the other consisted of extracts from Darwin's letters to Professor Henslow.); see also http://darwin-online.org.uk/contents. html#periodicals.

10 *Correspondence*, vol. 2, pp. 395–6; also available as letter 700 in *Darwin Corres-pondence Project* (http://www.darwin project.ac.uk).

11 Possibly the bookshop operated from 1832 until 1844 by former surveyor and explorer George Evans (Pike, 1966, pp. 359–60; see also Boswell, 1981).

12 *Correspondence*, vol. 4, pp. 229–31; also available as letter 1237 in *Darwin Corres-pondence Project* (http://www.darwin project.ac.uk).

13 *Ibid.*

14 *Correspondence*, vol. 4, pp. 368–70; also available as letter 1370 in *Darwin Correspondence Project* (http://www. darwinproject.ac.uk).

15 *Ibid.*

16 *Proceedings of the Linnean Society of New South Wales*, vol. 27 (part 3, no. 107, July to Sept. 1902): 344–5.

17 *Correspondence*, vol. 5, pp. 85–6; also available as letter 1477 in *Darwin Correspondence Project* (http://www. darwinproject.ac.uk).

18 *Correspondence*, vol. 5, pp. 163–4; also available as letter 1538 in *Darwin Correspondence Project* (http://www. darwinproject.ac.uk).

19 *Ibid.*

20 *Correspondence*, vol. 5, pp. 264–5; also available as letter 1637 in *Darwin Correspondence Project* (http://www. darwinproject.ac.uk).

21 In fact, with the exception of the index to the second volume of *Living Cirripedia*, which was not published until 1858, Darwin's barnacle books had all been published before he wrote the previous letter to Covington. The first volumes of each of *Living Cirripedia* and *Fossil Cirripedia* were actually published as early as 1851, and the second volumes of each work were published in 1854 (Freeman, 1977, pp. 66–8). Also, the Royal Society Medal was awarded to Darwin in 1853 (Clark, 1984, p. 92).

22 *Correspondence*, vol. 6, pp. 55–6; also available as letter 1840 in *Darwin Correspondence Project* (http://www. darwinproject.ac.uk).

23 *Correspondence*, vol. 6, pp. 345–6; also available as letter 2056 in *Darwin Correspondence Project* (http://www. darwinproject.ac.uk).

24 *Ibid.*

25 *Ibid.*

26 For the NSW Heritage listing, see http:// www.heritage.nsw.gov.au/07_subnav_04_ 1.cfm

27 The information on Covington's house was kindly supplied by the late Mrs O. Robertson, Mrs B. Ferguson and Mrs E. Beatty. Mrs Beatty has verified that the Forest Oak Inn was licensed in 1856, with Syms Covington as licensee.

28 *Correspondence*, vol. 7, pp. 95–6; also available as letter 2276 in *Darwin Correspondence Project* (http://www. darwinproject.ac.uk).

29 The main sources of the account of Darwin's activities presented in this and the following paragraphs are Oldroyd, 1980, ch. 7, and Clark, 1984, ch. 5. In passing, it should be noted that the date of Darwin's ninth letter to Covington, namely 18 May 1858, is strong evidence against the suggestion of Brooks, 1984, as cited by Clark, 1984, p. 105, that Wallace's letter arrived on 18 May, rather than as is generally thought, on 18 June. From the detailed and well-known account of Darwin's daily routine (as described by his son Francis in Darwin, 1887a, vol. 1, pp. 112–13, 118–19), it is clear that Darwin's mail always arrived and was opened in the morning, and that he wrote letters in the afternoon. This being so, it is incomprehensible that Darwin could have written such a serene letter to Covington just a few hours after receiving Wallace's manuscript.

30 Ear specialist.

31 *Correspondence*, vol. 7, p. 235; also available as letter 2400 in *Darwin Correspondence Project* (http://www.darwinproject.ac.uk).

32 Information on Martens has been obtained from the following sources: Gray, 1978; Keynes, 1979; Lindsay, 1968; McCulloch, 1984, pp. 773–4; Grainger, 1982, pp. 229–31; Pearce, 1979; and Anon., 1980.

33 *Heads of the People: an Illustrated Journal of Literature, Whims, and Oddities*, vol. 1, no. 20, Saturday 28 August, 1847, p. 145. Mitchell Library DSM/Q059/H.

34 *Correspondence*, vol. 9, p. 224; also available as letter 3222 in *Darwin Correspondence Project* (http://www.darwinproject.ac.uk).

35 Mitchell Library MSS 139/36X, part 2, p. 267. The whole letter comprises nine manuscript pages (263–72). Reproduced in *Correspondence*, vol. 9, pp. 318–20; also available as letter 3298 in *Darwin Correspondence Project* (http://www. darwinproject.ac.uk).

36 Darwin papers, Cambridge University Library, *N 10018 From Martens, Conrad 1862.01.20 DAR 171. Reproduced in *Correspondence*, vol. 10, pp. 35–6;

also available as letter 3398 in *Darwin Correspondence Project* (http://www.darwinproject.ac.uk).

37 The main sources for this brief summary of Wickham's life are Pike, 1967, p. 597; and Stokes, 1846.

38 These quotations are from Stokes, 1846, vol. 2, pp. 5–6.

39 Letters relating to the visit are in *Correspondence*, vol. 10, pp. 436–7, 443, 459, 532–3; also available as letters 3741, 3749, 3759 and 3809, respectively, in *Darwin Correspondence Project* (http://www.darwinproject.ac.uk). The actual date of the visit is recorded in Emma Darwin's diary (DAR242); see http://darwin-online.org.uk/EmmaDiaries.html

40 The case in favour of the claim is made by Thomson *et al.*, 1998; the case against is presented by Chambers, 2004, and Nicholls, 2006. Limited documentary research by the present authors supports the case against. We are grateful to Bethany Wilson for drawing our attention to the latter two articles.

41 The main sources for this section are Grainger, 1982, ch. 11; Pike, 1967, pp. 61–4; Gregson, 1907; and Walsh, 1967.

42 We are grateful to Ms L. Masson from the Stanton Library, North Sydney, for providing maps showing the exact location of 'Grantham'.

43 Clarke, 1856.

44 The main sources for this section are Pike, 1974, pp. 29–30; and Gregson, 1907.

45 King family papers, Mitchell Library FM4/6900, part B (ii), item 5, frames 326–9; also available as letter 1554a in *Darwin Correspondence Project* (http://www.darwinproject.ac.uk).

46 *Ibid*. We are grateful to Dr Tony Underwood and Professor Don Anderson for their assistance in deciphering Darwin's sentences about barnacles.

47 *Ibid*.

48 Cambridge University Library, Darwin papers, *N 17741 F 1862.09.16 DAR 169.

Reproduced in *Correspondence*, vol. 10, pp. 413–14; also available as letter 3741 in *Darwin Correspondence Project* (http://www.darwinproject.ac.uk).

49 *Ibid*.

50 *Ibid*.

51 King family papers, Mitchell Library FM4/6900, part B (ii), item 5, frames 330–3. Reproduced in *Correspondence*, vol. 10, pp. 532–3; also available as letter 3809 in *Darwin Correspondence Project* (http://www.darwinproject.ac.uk).

52 Cambridge University Library, Darwin papers, 1863.04.19 *R DAR 169. Reproduced in *Correspondence*, vol. 11, p. 329; also available as letter 4109 in *Darwin Correspondence Project* (http://www.darwinproject.ac.uk).

53 Cambridge University Library, Darwin papers, *N17743 1869.02.25 *R DAR 169. To be reproduced in a future volume of *Correspondence*, and, four years later, to become available as letter 6635 in *Darwin Correspondence Project* (http://www.darwinproject.ac.uk).

54 King family papers, Mitchell Library FM4/6900, part B (ii), item 5, frames 334–7. To be reproduced in a future volume of *Correspondence*, and, four years later, to become available as letter 6712a in *Darwin Correspondence Project* (http://www.darwinproject.ac.uk).

55 Cambridge University Library, Darwin papers, DAR 112.

56 Darwin, 1887, vol. 1, p. 223.

57 King's acceptance is recorded in a letter sent to Hallam Murray dated 27 Nov. 1890. We thank Ms V. Murray of John Murray, Publishers, for supplying a copy of the King–Murray correspondence.

58 On p. v of Barlow, 1933, Nora Barlow states that one of the drawings was used by Hallam Murray in the illustrated 1890 edition of the 1845 *Journal*, while Freeman, 1977, pp. 37, 42 comments that only some of the 1890 edition contained an additional plate with

two of King's sketches, and that this plate is not entered in the list of illustrations. In fact, a plate with two sketches was first published in the 1901 illustrated John Murray edition, where it was included in the list of illustrations. A revised plate was later incorporated (without inclusion in the list of illustrations, as noted by Freeman) into the 1913 reprint of the illustrated 1890 John Murray edition. In both editions, the plate faces page 1.

59 The original manuscript is in private possession, but a microfilm is held in the Mitchell Library: King family, P. P. King Papers 1806, 1826–1903; ZML MSS 3447; FM4/6900; B (iii) Reminiscences of Charles Darwin 1831–1836 by Philip Gidley King, the younger, Sent to W.A.H. Hallam Murray, October 17, 1892.

60 The main sources for this section are Mellersh, 1968; Scholefield, 1940, pp. 262–5; and Gribbin and Gribbin, 2003.

61 King family papers, Mitchell Library FM4/6900, part A (ii), item no. 3, frames 321–37.

62 Second editions are listed only when they involve substantial revision. With the exception of the first edition of *On the Movement and Habits of Climbing Plants*, which was published by Longman, Green, Longman, Roberts & Green and Williams & Norgate, London, all other volumes were published by John Murray, London. Details taken from Freeman, 1977.

63 The second edition (1860) was little more than a reprint of the first (1859) edition. In contrast, the third, fourth, fifth and sixth editions each involved substantial rewriting.

Bibliography

....

Adams, D. (ed.), 1952, *The Letters of Rachel Henning*, Sydney: Bulletin Newspaper Co.

Anon. n.d.(a), 'Historic Village: Battery Point', Hobart: Battery Point Village Promotion Association.

—— n.d.(b), 'The Hannibal Macarthurs', Pamphlet printed in England.

—— 1831–6, 'Logbook of the *Beagle*', Kew, Richmond, Surrey, England, Public Record Office, ADM51/3055:6228.

—— 1835a, [reference to building house at Vineyard] *The Colonist*, vol. I, no. 32, 6 August, p. 254, 2nd column.

—— 1835b, *The New South Wales Calendar and General Post Office Directory 1835*, Sydney: Stephens & Stokes.

—— 1836, *The New South Wales Calendar and General Post Office Directory 1836*, Sydney: Stephens & Stokes.

—— 1902, 'Notes and exhibits', *Proceedings of the Linnean Society of NSW*, vol. 27, part 3, pp. 344–5.

—— 1950, 'Bridge building in New South Wales, Part 1, The early stone bridges', *Main Roads*, vol. 16, no. 2, pp. 36–43.

—— 1964, *Priceless Heritage: Historic Buildings of Tasmania*, Hobart: Platypus Publications.

—— 1967, 'A Brief History of the Site of Rheem Australia Limited, Rydalmere, New South Wales', Sydney: Rheem Australia, Pamphlet.

—— 1980, *Map of the Municipality of North Sydney Showing 100 Historical Places, Including Historical Notes*, North Sydney: North Shore Historical Society.

—— 1983, *The Heritage of Tasmania: the Illustrated Register of the National Estate*, South Melbourne: Macmillan.

——2005, 'Heritage Assessment: Dunheved Precincts, St Marys Development, St Marys, N.S.W.' Marrickville: Casey & Lowe Pty Ltd.

Armstrong, P. 1985, *Charles Darwin in Western Australia*, Nedlands, WA: University of Western Australia Press.

Australia Division of National Mapping 1975, *Australia 1:250,000 Map Series Gazetteer*, Canberra: Australian Government Publishing Service, for the Department of Minerals and Energy.

Auwers, A. 1884, 'Some remarks on the chain of meridian distances, measured around the earth by H.M.S. "Beagle" between the years 1831 and 1836', *Monthly Notices of the Royal Astronomical Society*, vol. 64, no. 6, pp. 303–46.

Babington, C. C. 1841, 'Dytiscidae Darwinianae; or, descriptions of the species of *Dytiscidae* collected by Charles Darwin, Esq., M.A., Sec. G.S. &c., in South America and Australia, during his voyage in H.M.S. Beagle', *Transactions of the Entomological Society of London*, vol. 3, no. 1, pp. 1–17.

....

Banks, M. R. 1971, 'A Darwin manuscript on Hobart Town', *Papers and Proceedings of the Royal Society of Tasmania*, vol. 105, pp. 5–19.

Banks, M. R. and D. Leaman 1999, 'Charles Darwin's field notes on the geology of Hobart town — a modern appraisal', *Papers and Proceedings of the Royal Society of Tasmania*, vol. 133, pp. 29–50.

Barca, M. (ed.), 1979, *Advice to a Young Lady in the Colonies*, Collingwood, Melbourne: Greenhouse Publications.

Barlow, N. (ed.), 1933, *Charles Darwin's Diary of the Voyage of H.M.S. Beagle*, Cambridge: Cambridge University Press.

—— (ed.), 1945, *Charles Darwin and the Voyage of the Beagle*, London: Pilot Press.

—— 1950, 'Darwin and the *Beagle*'s artist', *Geographical Magazine*, vol. 23, pp. 277–81.

—— (ed.), 1958, *The Autobiography of Charles Darwin 1809–1882, with the Original Omissions Restored*, London: Collins.

Barrett, P. H. (ed.), 1977, *The Collected Papers of Charles Darwin*, vol. 2, Chicago: University of Chicago Press.

Barrett, P. H. and R. B. Freeman (eds.), 1986, *The Works of Charles Darwin*, London: Pickering and Chatto.

Bates, G. 1981, *Gardner's Inn, Blackheath, Blue Mountains, NSW: Sesquicentenary 1831–1981*, Blackheath: The Author.

Beard, J. S., A. R. Chapman and P. Gioia 2000, 'Species richness and endemism in the Western Australian flora', *Journal of Biogeography*, vol. 27, pp. 1257–68.

Bedford, R. 1954, *Think of Stephen*, Sydney: Angus and Robertson.

Bell, T. 1842–43, *The Zoology of the Voyage of H.M.S. Beagle, under the command of Captain FitzRoy, R.N., during the Years 1832 to 1836. Part V. Reptiles (and Amphibia) [in 2 numbers]*. Edited by C. R. Darwin, London: Smith, Elder and Co.

Bischoff, J. 1832, *Sketch of the History of Van Diemen's Land*, London: John Richardson.

Blakers, M., S. J. J. F. Davies and P. N. Reilly 1984, *The Atlas of Australian Birds*, Melbourne: Melbourne University Press.

Blumenbach, J. F. 1800, 'Über das Schnabelthier (Ornithorhynchus paradoxus) ein neuentdecktes Geschlecht von Säugthieren des fünften Welttheils', *Magazin für den Neuesten Zustand der Naturkunde*, vol. 2, pp. 205–214.

Bornemissza, G. F. 1983, 'Darwin and the Tasmanian dung beetles', *Tasmanian Naturalist*, vol. 75, pp. 2–4.

Boswell, A. 1981, *Annabella Boswell's Journal: An Account of Early Port Macquarie*, Sydney: Angus and Robertson.

Branagan, D. F. 1985, 'The Blue Mountains – A Personal Perspective' in P. Stanbury and L. Bushell (eds.), *The Blue Mountains: Grand Adventure for All*, Sydney: Macleay Museum, University of Sydney, pp. 1–13.

Branagan, D. F. and G. H. Packham 2000, *Field Geology of New South Wales*, Sydney: NSW Department of Mineral Resources.

Broadbent, J. and M. Dupain 1978, *The Golden Decade of Australian Architecture – The Work of John Verge*, Sydney: David Ell Press.

Brooks, J. L. 1984, *Just Before the Origin: Alfred Russel Wallace's Theory of Evolution*, New York: Columbia University Press.

Browne, J. 1995, *Charles Darwin: Voyaging*, London: Jonathan Cape.

—— 2002, *Charles Darwin: The Power of Place*, London: Jonathan Cape.

Bruton, E. 1968, *Clocks and Watches*, Feltham, Sussex: Paul Hamlyn.

Burkhardt, F. and S. Smith (eds.), 1985, *Calendar of Correspondence of Charles Darwin, 1821–1882*, New York: Garland.

Burstyn, H. L. 1975, 'If Darwin wasn't the Beagle's naturalist, why was he on board?' *British Journal for the History of Science*, vol. 8, no. 28, pp. 62–9.

Carrington, H. 1939, *Life of Captain Cook*, London: Sidgwick and Jackson.

Chambers, P. 2004, 'The origin of Harriet', *New Scientist*, vol. 183, no. 2464, p. 38.

[Chambers, R.] 1844, *Vestiges of the Natural History of Creation*, London: John Churchill.

Chessell, G. 2005, *Richard Spencer: Napoleonic War Naval Hero and Australian Pioneer*, Crawley, WA: University of Western Australia Press.

Chorley, R. J., A. J. Dunn and R. P. Beckinsale 1964, *The History of the Study of Landforms or the Development of Geomorphology. Volume One: Geomorphology Before Davis*, London: Methuen and John Wiley.

Clark, R. W. 1984, *The Survival of Charles Darwin: A Biography of a Man and an Idea*, London: Weidenfeld and Nicolson.

Clarke, W. B. 1856, *The dead which are blessed, A Sermon Preached in the Church of St Thomas, Willoughby, NSW on Sunday, 2nd March, 1856, the Day after the Funeral of the Late Rear-Admiral Philip [sic] Parker King, MC, Printed by Particular Desire, for Private Distribution Only*, Sydney: Joseph Cook & Co.

Clifford, B. 1961, 'Subiaco no more', *Walkabout*, vol. 27, no. 11, pp. 26–7.

Cogger, H. G. 1983, *Reptiles and Amphibians of Australia*, 3rd edn, Sydney: Reed.

Cole, J. 1980, 'The Vineyard that dies – Women of Subiaco', in P. Thompson and S. Yorke (eds.), *Lives Obscurely Great*, Sydney: Society of Women Writers [Australia] NSW Branch.

—— (ed.), 1983, *Parramatta River Notebook*, Sydney: Kangaroo Press.

Collins, D. 1802, *An Account of the English Colony in New South Wales*, vol. 2, London: Cadell and Davies.

Covington, S. 1831–6, Diary of visit to Australia, Linnean Society of NSW Records, 1826–1941, VII, Papers of Syms Covington, 1831–6, 1839, Mitchell Library MSS 2009/108 item 5. Available at http://www.asap.unimelb.edu.au/bsparcs/covingto/contents.htm

Daley, C. 1938, 'Charles Darwin and Australia', *Victorian Historical Magazine*, vol. 17, no. 2, pp. 64–70.

Darwin, C. R. (ed.), 1838–1845, *The Zoology of the Voyage of H.M.S. Beagle, under the command of Captain FitzRoy, R.N., during the Years 1832 to 1836*, London: Smith, Elder and Co.

—— 1839, *Journal and Remarks, 1832-1836, Narrative of the Surveying Voyages of His Majesty's Ships Adventure and Beagle between the years 1826 and 1836, describing their Examination of the Southern Shores of South America, and the Beagle's Circumnavigation of the Globe (in three volumes, plus an appendix to volume II)*, Edited by R. FitzRoy, vol. III, London: Henry Colburn.

—— 1839, *Journal of Researches into the Geology and Natural History of the Various Countries visited by H.M.S. Beagle, under the command of Captain FitzRoy, R.N., from 1832 to 1836*, London: Henry Colburn.

—— 1842, *The Structure and Distribution of Coral Reefs. Being the first part of the Geology of the Voyage of the Beagle, under the command of Capt FitzRoy, R.N., during the Years 1832 to 1836*, London: Smith, Elder and Co.

—— 1844a. *Geological Observations on the Volcanic Islands visited during the Voyage of H.M.S. Beagle, together with some brief Notices of the Geology of Australia and the Cape of Good Hope. Being the second part of the Geology of the Voyage of the Beagle, under the command of Capt FitzRoy, R.N., during the years 1832 to 1836.* London: Smith, Elder and Co.

—— 1844b. 'Brief descriptions of several terrestrial Planariae, and of some remarkable marine species, with an account of their habits', *Annals and Magazine of Natural History, including Zoology, Botany, and Geology,* vol. 14, pp. 241–51.

—— 1845. *Journal of Researches into the Natural History and Geology of the Countries visited during the Voyage of H.M.S. Beagle round the World, under the command of Capt. FitzRoy, R.N., 2nd edn, corrected, with additions.* London: John Murray.

—— 1846. *Geological Observations on South America. Being the third part of the Geology of the Voyage of the Beagle, under the command of Capt FitzRoy, R.N., during the years 1832 to 1836.* London: Smith, Elder and Co.

—— 1859. *On the Origin of Species by Means of Natural Selection, or the Preservation of Favoured Races in the Struggle for Life,* 1st edn, London: John Murray.

—— 1860. *On the Origin of Species by Means of Natural Selection, or the Preservation of Favoured Races in the Struggle for Life,* 2nd edn, London: John Murray.

—— 1861. *On the Origin of Species by Means of Natural Selection, or the Preservation of Favoured Races in the Struggle for Life,* 3rd edn, London: John Murray.

—— 1862. *On the Various Contrivances by Which British and Foreign Orchids are Fertilised by Insects, and on the Good Effects of Intercrossing,* 1st edn. London: John Murray.

—— 1865. *On the Movements and Habits of Climbing Plants,* 1st edn, London: Longman, Green, Longman, Roberts & Green and Williams & Norgate.

—— 1866. *On the Origin of Species by Means of Natural Selection, or the Preservation of Favoured Races in the Struggle for Life,* 4th edn, London: John Murray.

—— 1868. *The Variation of Animals and Plants under Domestication,* 2 vols, 1st edn. London: John Murray.

—— 1869. *On the Origin of Species by Means of Natural Selection, or the Preservation of Favoured Races in the Struggle for Life,* 5th edn, London: John Murray.

—— 1871. *The Descent of Man, and Selection in Relation to Sex,* 2 vols, 1st edn, London: John Murray.

—— 1872. *On the Origin of Species by Means of Natural Selection, or the Preservation of Favoured Races in the Struggle for Life,* 6th edn, London: John Murray.

—— 1872. *The Expression of the Emotions in Man and Animals,* London: John Murray.

—— 1874. *The Descent of Man, and Selection in Relation to Sex,* in one volume, 2nd edn, London: John Murray.

—— 1875. *Insectivorous Plants,* London: John Murray.

—— 1875. *On the Movements and Habits of Climbing Plants,* 2nd edn, London: John Murray.

—— 1875. *The Variation of Animals and Plants under Domestication,* 2 vols, 2nd edn, London: John Murray.

—— 1876. *The Effects of Cross and Self Fertilisation in the Vegetable Kingdom,* London: John Murray.

—— 1877. *On the Various Contrivances by Which British and Foreign Orchids are Fertilised by Insects, and on the Good Effects of Intercrossing,* 2nd edn, London: John Murray.

—— 1877. *The Different Forms of Flowers on Plants of the Same Species,* London: John Murray.

—— 1881. *The Formation of Vegetable Mould, through the Action of Worms, with Observations on Their Habits,* London: John Murray.

Darwin, C. R. and F. Darwin 1880, *The Power of Movement in Plants*, London: John Murray.

Darwin, F. (ed.), 1887a, *The Life and Letters of Charles Darwin, including an Autobiographical Chapter*, 2 vols, London: John Murray.

—— 1887b, 'Reminiscences of my father's everyday life', in F. Darwin (ed.), *The Life and Letters of Charles Darwin*, London: John Murray, pp. 112–13, 118–19.

—— (ed.), 1909, *The Foundations of the Origin of Species. Two Essays written in 1842 and 1844*, Cambridge: Cambridge University Press.

—— 1912, 'FitzRoy and Darwin', *Nature*, vol. 88, no. 2208, pp. 547–8.

Dawson, T. J. 1983, *Monotremes and Marsupials: The Other Mammals*, London: Edward Arnold.

de Beer, G. 1959, 'Some unpublished letters of Charles Darwin', *Notes and Records of the Royal Society of London*, vol. 14, pp. 12–66.

de Falbe, E. 1860–94, Letters and recollections of Emmeline de Falbe, 1860–94, Macarthur Family papers, vol. 81, A2977, Mitchell Library, Sydney.

—— 1954a, 'An early-day aristocrat', *The Bulletin*, 10 March, pp. 24, 25, 34.

—— 1954b, 'An early-day aristocrat', *The Bulletin*, 17 March, pp. 24, 25, 31.

de Falbe, J. (ed.), 1988, *My Dear Miss Macarthur: the recollections of Emmeline Maria Macarthur (1828–1911)*, Kenthust, NSW: Kangaroo Press.

de Vries-Evans, S. 1993, *Conrad Martens on the Beagle and in Australia*, Brisbane: Pandanus Press.

Desmond, A. and J. Moore 1991, *Darwin*, London: Michael Joseph.

Dickman, C. R. 2007, *A Fragile Balance: the Extraordinary Story of Australian Marsupials*, Fisherman's Bend, Vic: Craftsman House.

Dietz, R. S. 1978, 'IFOs (Identified Flying Objects)', *Sea Frontiers*, vol. 24, pp. 341–6.

Douglass, N. 1985, 'Inn at Weatherboard (now Wentworth Falls): Background History of the Inn and Its Site, a report prepared for the Blue Mountains City Council and the Heritage Council of NSW, 31st March 1985', Katoomba: Blue Mountains City Council and the Heritage Council of NSW.

Dow, G. 1974, *Samuel Terry: The Botany Bay Rothschild*, Sydney: Sydney University Press.

Dundas, D. 1979, *The Art of Conrad Martens*, Melbourne: Macmillan.

Dupain, M. and M. Herman 1963, *Georgian Architecture in Australia*, Sydney: Ure Smith.

Earle, A. 1832, *A Journal of a Residence in Tristan d'Acunha, an Island Situated between South America and the Cape of Good Hope*, London: Longman, Rees, Orme, Brown, Green, and Longman.

Eldershaw, M. B. 1973, *The Life and Times of Captain John Piper*, Sydney: Ure Smith, in association with the National Trust of Australia [NSW].

Ellis, E. 1994, *Conrad Martens: Life and Art*, Sydney: State Library of NSW Press.

Erickson, R., A. S. George, N. G. Marchant and M. K. Morcombe, (eds.), 1973, *Flowers and Plants of Western Australia*, Sydney: Reed.

Evely, D.A. 2003, 'Alexander Macleay and his contribution to colonial New South Wales', PhD thesis, Sydney: University of Sydney.

Ferguson, B. J. 1971, *Syms Covington of Pambula*, Bega, NSW: Imlay District Historical Society.

FitzRoy, R. (ed.), 1839a, *Narrative of the Surveying Voyages of His Majesty's Ships Adventure and Beagle between the years 1826 and 1836, describing their Examination of the Southern Shores of South America, and the Beagle's Circumnavigation of the Globe (in three volumes, plus an appendix to volume II)*, London: Henry Colburn.

—— 1839b, *Proceedings of the Second Expedition, 1831–1836, under the command of Captain Robert Fitz-Roy, R.N., Narrative of the Surveying Voyages of His Majesty's Ships Adventure and Beagle between the years 1826 and 1836, describing their Examination of the Southern*

Shores of South America, and the Beagle's Circumnavigation of the Globe (in three volumes, plus an appendix to volume II). Edited by R. FitzRoy, vol. II. London: Henry Colburn.

—— 1863, *Weather Book: a Manual of Practical Meteorology*, London: Longman, Green, Longman, Roberts & Green.

FitzRoy, R. and C. R. Darwin 1836, 'A letter, containing remarks on the moral state of Tahiti, New Zealand &c.' *South African Christian Recorder*, vol. 2, no. 4, pp. 221–38.

Fleay, D. 1980, *Paradoxical Platypus*, Milton, Qld: Jacaranda Press.

Forshaw, J. M. and W. T. Cooper 1978, *Parrots of the World*, 2nd edn, Melbourne: Lansdowne.

—— 1981, *Australian Parrots*, 2nd edn, Melbourne: Lansdowne.

Frankland, G. 1836, 'A notice on Maria Island on the east coast of Van Diemen's Land, south latitude 42°44', east longitude 148°8'', *Proceedings of the Geological Society*, vol. 2, p. 415 only.

Freeman, R. B. 1977, *The Works of Charles Darwin: An Annotated Bibliographical Handlist*, 2nd edn, Folkestone, Kent: Dawson.

—— 1978, *Charles Darwin: a Companion*, Folkestone, Kent: Dawson.

Friendly, A. 1977, *Beaufort of the Admiralty: the Life of Sir Francis Beaufort, 1774–1857*, London: Hutchinson.

Garden, D. S. 1977, *Albany: A Panorama of the Sound from 1827*, West Melbourne: Nelson.

Goodwin, D. 1977, *Crows of the World*, London and Brisbane: British Museum [Natural History] and University of Queensland Press.

Gould, J. 1983, *The Mammals of Australia 1845–63, with modern notes by Joan M. Dixon*, South Melbourne: Macmillan.

Gould, J. and G. R. Gray 1838–1841, *The Zoology of the Voyage of H.M.S. Beagle, under the command of Captain FitzRoy, R.N., during the years 1832 to 1836. Part II. Birds [in 4 numbers]*, Edited by C. R. Darwin, London: Smith, Elder and Co.

Grainger, E. 1982, *The Remarkable Reverend Clarke*, Melbourne: Oxford University Press.

Gray, J. 1978, 'Introduction to the Queensland Art Gallery Catalogue June 1976' in *Catalogue of the Conrad Martens Centenary Exhibition 24th May – 23rd July 1978, S.H. Erwin Museum and Art Gallery*, Sydney: The National Trust of Australia (New South Wales).

Gregson, J. 1907, *The Australian Agricultural Company, 1824–1875*, Sydney: Angus and Robertson.

Gribbin, J. and M. Gribbin 2003, *FitzRoy: the Remarkable Story of Darwin's Captain and the Invention of the Weather Forecast*, London: Review Books.

Gruber, J. W. 1969, 'Who was the *Beagle*'s naturalist?' *British Journal for the History of Science*, vol. 4, no. 15, pp. 266–82.

Hackforth-Jones, J. 1980, *Augustus Earle, Travel Artist*, Canberra: National Library of Australia.

Hawkins, R. 1980, *Wallerawang: Some Aspects of Its History; 1823–1979*, Wallerawang: Wallerawang Public School.

Hazlewood, N. 2000, *Savage: the Life and Times of Jemmy Button*, London: Hodder and Stoughton.

Herman, M. 1954, *The Early Australian Architects and their Work*, Sydney: Angus and Robertson.

Hordern, M. C., 1989, *Mariners are Warned!: John Lort Stokes and H.M.S. Beagle in Australia 1837–1843*, Melbourne: Miegunyah Press.

—— 1997, *King of the Australian Coast: the Work of Phillip Parker King in the Mermaid and Bathurst 1817–1822*, Melbourne: Miegunyah Press.

Horner, B. E. and J. M. Taylor 1965, 'Systematic relationships among Rattus in southern Australia: evidence from cross-breeding experiments', *Wildlife Research*, vol. 10, no. 1, pp. 101–109.

Hügel, C. von 1994, *New Holland Journal: November 1833 – October 1934*, translated and edited by D. Clark, Melbourne: Miegunyah Press.

Hyett, J. and N. Shaw 1980, *Australian Mammals*, Melbourne: Nelson.

Jack, R. I. 1987, 'Andrew Brown, laird of Cooerwull', *Journal of the Royal Australian Historical Society*, vol. 73, no. 3, pp. 173–86.

Jenyns, L. 1840–1842, *The Zoology of the Voyage of H.M.S. Beagle, under the command of Captain FitzRoy, R.N., during the years 1832 to 1836. Part IV. Fish [in 4 numbers]*. Edited by C. R. Darwin, London: Smith, Elder and Co.

Johnson, L. 1984, *More than a House: The Old Farm on Strawberry Hill*, Albany: Old Farm Strawberry Hill Management Committee.

Keynes, R. D. 1979, *The Beagle Record*, Cambridge: Cambridge University Press.

—— (ed.), 1988, *Charles Darwin's Beagle Diary*, Cambridge: Cambridge University Press.

King, P. G. n.d., Memoirs of Governor King, Admiral King and an autobiography by Philip Gidley King, the younger, C770, Mitchell Library, Sydney.

—— 1891, *Comments on Cook's log (H.M.S. Endeavour, 1770): with extracts, charts and sketches*, Sydney: George Stephen Chapman, Acting Government Printer.

—— 1892, Reminiscences re Charles Darwin 1831–1836. Sent to W. A. H. Hallam Murray October 17, 1892, frames 339–46, FM4/6900, Mitchell Library, Sydney.

—— 1894, Autobiography 1894, frame 320, FM4/6900, Mitchell Library, Sydney.

King, P. P. 1839, *Proceedings of the First Expedition, 1826–1830, under the command of Captain P. Parker King, R.N., F.R.S., Narrative of the Surveying Voyages of His Majesty's Ships Adventure and Beagle between the years 1826 and 1836, describing their Examination of the Southern Shores of South America, and the Beagle's Circumnavigation of the Globe (in three volumes, plus an appendix to volume II)*. Edited by R. FitzRoy, vol. I, London: Henry Colburn.

King, P. P. and W. J. Broderip 1832–4, 'Description of the Cirrhipeda, Conchifera and Mollusca in a collection formed by the Officers of H.M.S. Adventure and Beagle employed between the years 1826 and 1830 in surveying the Southern Coasts of South America, including the Straits of Magalhaens and the Coast of Tierra del Fuego', *Zoological Journal*, vol. 5, pp. 332–49.

Kirby, W. and W. Spence 1815–26, *An Introduction to Entomology*, 4 vols, London: Longman, Hurst, Rees, Orme & Brown.

Lea, A. M. 1926, 'On some Australian Coleoptera collected by Charles Darwin during the voyage of the "Beagle"', *Transactions of the Entomological Society of London*, vol. 74, no. 2, pp. 279–88.

Lindsay, L. 1968, *Conrad Martens: The Man and his Art*, 2nd edn, Sydney: Angus and Robertson.

Linssen, E. T. 1959, *Beetles of the British Isles, First Series*, London: Frederick Warne.

Low, J. 1983, *The Bridge at Emu Pass*, Springwood: Blue Mountains City Library.

Lucas, C. 1976, 'Reply to "Entrance doors of 'Subiaco'"', *Newsletter of the Royal Australian Historical Society*, September, p. 2.

Lukis, M. n.d., *The Old Farm, Strawberry Hill, Albany, Western Australia*, Perth: National Trust of Australia [WA].

Mackaness, G. (ed.), 1965, *Fourteen Journeys Over the Blue Mountains of New South Wales 1813–1841*, Sydney: Horwitz-Grahame.

Malthus, T. R. 1798, *An Essay on the Principle of Population, as it Affects the Future Improvement of Society; With Remarks on the Speculations of W. Godwin, M. Condorcet and Other Writers*, London: J. Johnson.

Marshall, A. J. 1970, *Darwin and Huxley in Australia*, Sydney: Hodder and Stoughton.

McCormick, E. H. (ed.), 1966, *Narrative of a Residence in New Zealand; Journal of a Residence in Tristan da Cunha; by Augustus Earle*. Oxford: Clarendon Press.

McCulloch, A. (ed.), 1984, *Encyclopaedia of Australian Art, Volume Two, L–Z*, Melbourne: Hutchinson of Australia.

McDonald, R. 1998, *Mr Darwin's Shooter*. Sydney: Knopf.

McKenzie, E. J. 1966, Lithgow and District ([Notes] Prepared by EJM for a tour of Lithgow and district by members of the Royal Australian Historical Society, Saturday, May 28, 1966), Q991.6/13, Mitchell Library, Sydney.

Mellersh, H. E. L. 1968, *FitzRoy of the Beagle*, New York: Mason and Lipscomb.

Mowle, L. M. 1978, *A Genealogical History of Pioneer Families of Australia*, Sydney: Rigby.

Moyal, A. 1986, *A Bright and Savage Land*, Sydney: Collins.

—— 2001, *Platypus*, Sydney: Allen and Unwin.

—— 2003, *The Web of Science: the Scientific Correspondence of the Rev. W.B. Clarke, Australia's Pioneer Geologist*, in two volumes, Melbourne: Australian Scholarly Publishing.

Mullins, B. and D. Baglin 1978, *Western Australian Wildflowers in Colour*, Sydney: Reed.

Nairn, B. (ed.), 1976, Entry for A. Stephen in *Australian Dictionary of Biography 1851–1890, R–Z*, Melbourne: Melbourne University Press.

New South Wales Department of Main Roads 1949, 'The Great Western Highway: a romance of early road building', *Main Roads*, vol. 15, no. 1, pp. 6–15.

Newell, H. H. 1938, 'Road engineering and its development in Australia, 1788–1938', *Journal of the Institution of Engineers, Australia*, vol. 10, nos. 2,3, pp. 41–70, 97–106.

Nicholas, F. W. 2008, 'What Darwin actually saw in Sydney in 1836', *Australian Zoologist*, (in press).

Nicholls, H. 2006, 'Tall tales and tortoises: from Newton's apple to Darwin's finches, the history of science is littered with myths and half-truths. Should we care', *New Scientist*, vol. 191, no. 2560, p. 21.

Nicholson, I. H., 1983, *Shipping Arrivals and Departures, Tasmania, Volume 1, 1803–1833*, Canberra: Roebuck Society.

Norman, L. 1938, *Pioneer Shipping of Tasmania: whaling, sealing, piracy, shipwrecks, etc. in early Tasmania*, Hobart: J. Walch & Sons.

Oldroyd, D. R. 1980, *Darwinian Impacts*, Sydney: University of New South Wales Press.

O'May, H. 1959, *Hobart River Craft and Sealers of Bass Strait*, Hobart: Government Printer.

O'Shaughnessy, E. W. 1835, *Australian Almanack & Directory 1835*, Pitt Street, Sydney: John Innes.

Owen, R. 1838–1840, *The Zoology of the Voyage of H.M.S. Beagle, under the command of Captain FitzRoy, R.N., during the years 1832 to 1836. Part I. Fossil Mammalia [in 4 numbers]*, Edited by C. R. Darwin, London: Smith, Elder and Co.

Pearce, B. (ed.), 1979, *Conrad Martens: The H. W. B. Chester Memorial Collection (Catalogue for the Exhibition held at the Art Gallery of New South Wales, 22 Dec. 1979 to 10 Feb. 1980)*, Sydney: Art Gallery of New South Wales.

Phillip, A. 1789, *The Voyage of Governor Phillip to Botany Bay, With an Account of the Establishment of the Colonies of Port Jackson and Norfolk Island*, London: John Stockdale.

Pickard, W. J. 1976, 'Entrance doors of "Subiaco"', *Newsletter of the Royal Australian Historical Society*, August, p. 4.

Pickett, J. W. and J. D. Alder 1997, *Layers of time: the Blue Mountains and their Geology*, Sydney: NSW Department of Mineral Resources.

Pike, D. (ed.), 1966, Entries for G. Evans and G. Frankland in *Australian Dictionary of Biography 1788–1850, A–H*, Melbourne: Melbourne University Press.

—— (ed.), 1967, Entries for P. P. King, H. Macarthur and J. C. Wickham in *Australian Dictionary of Biography 1788–1850, I–Z*, Melbourne: Melbourne University Press.

—— (ed.), 1974, Entry for P. G. King in *Australian Dictionary of Biography 1851–1890, K–Q*, Melbourne: Melbourne University Press.

Raper, H. 1914, *The Practice of Navigation and Nautical Astronomy*, 20th edn, London: J. D. Potter.

Reed, T. T. 1978, *Historic Churches of Australia*, Melbourne: Macmillan.

Reid, G. 1988, *From Dusk Till Dawn. A History of Australian Lighthouses*, Sydney: Commonwealth Department of Transport and Communications and Macmillan.

Rickwood, P. C. and D. J. West (eds.), 2005, *Blackheath: Today from Yesterday*, Blackheath, NSW: WriteLight, for the Rotary Club of Blackheath.

Ride, W. D. L. and E. Fry 1970, *A Guide to the Native Mammals of Australia*, Melbourne: Oxford University Press.

Robertson, E. G. 1970, *Early Buildings of Southern Tasmania*, vol. 1, Melbourne: Georgian House.

Robson, L. 1983, *A History of Tasmania, Volume 1: Van Diemen's Land from the Earliest Times to 1855*, Melbourne: Oxford University Press.

—— 1985, *A Short History of Tasmania*, Melbourne: Oxford University Press.

Rowntree, A. 1951, *Battery Point Today and Yesterday*, Hobart: Tasmanian Department of Education.

Roxburgh, R. 1974, *Early Colonial Houses of New South Wales*, Sydney: Ure Smith.

Ryan, L. 1981, *The Aboriginal Tasmanians*, St. Lucia, Qld: University of Queensland Press.

Saunders, W. W. 1843, 'Descriptions of new Australian *Chrysomelidae* allied to *Cryptocephalus*', *Annals and Magazine of Natural History*, vol. 11, April, p. 317 only.

Scholefield, G. H. (ed.), 1940, *A Dictionary of New Zealand Biography, Volume I, A–L*, Wellington, NZ: Department of Internal Affairs.

Semeniuk, V. and T. D. Meagher 1981, 'Calcrete in quaternary coastal dunes in southwestern Australia: a capillary-rise phenomenon associated with plants', *Journal of Sedimentary Petrology*, vol. 51, no. 1, pp. 47–68.

Shaw, G. 1799, 'Platypus anatinus. The Duck-billed Platypus', *The Naturalists' Miscellany*, London: F. P. Nodder & Co., no. 10, 7 unnumbered pages plus plates 385–6.

Smith, K. G. V. 1987, 'Darwin's insects: Charles Darwin's entomological notes', *Bulletin of the British Museum (Natural History) (Historical Series)*, vol. 14, pp. 1–143.

—— 1996, 'Supplementary notes on Darwin's insects', *Archives of Natural History*, vol. 23, no. 2, pp. 279–86.

Sobel, D. 1995, *Longitude*, New York: Walker.

Stanbury, D. (ed.), 1977, *A Narrative of the Voyage of H.M.S. Beagle*, London: Folio Society.

Stokes, J. L. 1846, *Discoveries in Australia; with an Account of the Coasts and Rivers Explored and Surveyed during the Voyage of H.M.S. Beagle, in the Years 1837–38–39–40–41–42–43*, 2 vols, London: T. and W. Boone.

Strahan, R. (ed.), 1983, *The Complete Book of Australian Mammals*, Sydney: Angus and Robertson.

Strzelecki, P. E. 1845, *Physical Description of New South Wales and Van Diemen's Land*, London: Longman, Brown, Green, & Longmans.

Sutherland, F.L. and P. Wellman 1986, 'Potassium-argon ages of tertiary volcanic rocks, Tasmania', *Papers and Proceedings of the Royal Society of Tasmania*, vol. 120, pp. 77–86.

Syme, P. 1821. *Werner's Nomenclature of Colours with Additions, Arranged So As To Render It Highly Useful to the Arts and Sciences, Particularly Zoology, Botany, Chemistry, Minerology, and Morbid Anatomy*. 2nd edn. Edinburgh: Blackwood.

Thomson, S., S. Irwin and T. Irwin 1998. 'Harriet the Galapagos tortoise: disclosing one and a half centuries of history', *Reptilia*, vol. 2, March–April, pp. 48–51.

Thorp, W. 1985. 'Archaeological Investigation: Weatherboard Inn Site, Wentworth Falls, a report prepared for the Blue Mountains City Council, April 1985', Heritage Council Library, Katoomba. Q944.45/THO.

Tschudi, J. J. v. 1838. *Classification der Batrachier, mit Berücksichtigung der fossilen Tiere dieser Abteilung der Reptilien*. Neuchâtel: Erück Petitpierre.

Turnbull, C. 1948. *Black War: The Extermination of the Tasmanian Aborigines*. Melbourne: F. W. Chesire.

Vallance, T. G. 1975. 'Presidential Address: Origins of Australian geology', *Proceedings of the Linnean Society of New South Wales*, vol. 100, no. 1, pp. 13–43.

van Dyck, S. and R. Strahan (eds.). 2008. *The Mammals of Australia*, 3rd edn. Sydney: New Holland Publishers.

Verge, W. G. 1962. *John Verge, Early Australian Architect: His Ledger & His Clients*. Sydney: Wentworth.

Walsh, D. (ed.), 1967. *The Admiral's Wife: Mrs Phillip Parker King: A Selection of Letters 1817–56*. Melbourne: Hawthorn Press.

Waterhouse, G. R. 1838. 'Descriptions of some of the insects brought to this country by C. Darwin', *Transactions of the Entomological Society of London*, vol. 2, no. 2, pp. 131–5.

—— 1838–1839. *The Zoology of the Voyage of H.M.S. Beagle, under the command of Captain FitzRoy, R.N., during the years 1832 to 1836. Part II. Living Mammalia [in 4 numbers]*. Edited by C. R. Darwin. London: Smith, Elder and Co.

—— 1839, 'Descriptions of some new species of exotic insects', *Transactions of the Entomological Society of London*, vol. 2, no. 3, pp. 188–96, plus plate 17.

—— 1846, *A natural history of the Mammalia, vol. 1. Marsupiata or pouched animals*. London: Hippolyte Baillière.

Wheeler, J., N. G. Marchant and M. Lewington 2002. *Flora of the South West – Bunbury, Augusta, Denmark*. 2 vols. Canberra and Perth: Australian Biological Resources Study, and Western Australian Department of Conservation and Land Management.

Williams, S. 1982. *Pass of Victoria Sesquicentenary 1832–1982*, Mount Victoria, NSW: Mount Victoria and District Historical Society.

Wilson, A. 1917–19. 'Subiaco Benedictine monastery', *Journal of the Royal Australian Historical Society*, vol. 4, pp. 385–7.

Wilson, W. H. 1924. *Old Colonial Architecture in NSW and Tasmania*. Sydney: The author, at the Union House.

Winchester, F. 1972, James Walker of Wallerowang, [paper] presented to the Lithgow District Historical Society on 11th November 1972, Lithgow District Historial Society, Paper No. 11, 1–8 pp., Lithgow.

Wrigley, J. W. and M. Fagg 1983. *Australian Native Plants*, 2nd edn. Sydney: Collins.

INDEX

....

....
254